D1287414

Mixed
Analog-Digital VLSI
Devices and
Technology

Other Reference Books of Interest by McGraw-Hill

Handbooks

BENSON · *Television Engineering Handbook*

CHEN · *Computer Engineering Handbook*

COOMBS · *Printed Circuits Handbook*

DI GIACOMO · *Digital Bus Handbook*

DI GIACOMO · *VLSI Handbook*

FINK AND CHRISTIANSEN · *Electronics Engineers' Handbook*

HARPER · *Electronic Packaging and Interconnection Handbook*

JURAN AND GRYNA · *Juran's Quality Control Handbook*

RORABAUGH · *Digital Filter Designer's Handbook*

SERGENT AND HARPER · *Hybrid Microelectronic Handbook*

TUMA · *Engineering Mathematics Handbook*

WAYNANT · *Electro-Optics Handbook*

WILLIAMS AND TAYLOR · *Electronic Filter Design Handbook*

Other

ANTOGNETTI · *Power Integrated Circuits*

ANTOGNETTI AND MASSOBRIO · *Semiconductor Device Modeling with SPICE*

BEST · *Phase-Locked Loops*

BUCHANAN · *CMOS / TTL Digital Systems Design*

BUCHANAN · *BiCMOS / CMOS Systems Design*

BYERS · *Printed Circuit Board Design with Microcomputers*

ELLIOTT · *Integrated Circuits Fabrication Technology*

HECHT · *The Laser Guidebook*

KIELKOWSKI · *Inside SPICE*

SMITH · *Thin-Film Deposition*

SZE · *VLSI Technology*

TSUI · *LSI / VLSI Testability Design*

WOBSCHALL · *Circuit Design for Electronic Instrumentation*

WYATT · *Electro-Optical System Design*

To order or receive additional information on these or any other McGraw-Hill titles, please call 1-800-822-8158 in the United States. In other countries, contact your local McGraw-Hill representative.   **BC15XXA**

Mixed Analog-Digital VLSI Devices and Technology

An Introduction

Yannis Tsividis
Columbia University

RECEIVED
DEC 27 1996
MSU - LIBRARY

McGraw-Hill

New York   San Francisco   Washington, D.C.   Auckland   Bogotá
Caracas   Lisbon   London   Madrid   Mexico City   Milan
Montreal   New Delhi   San Juan   Singapore
Sydney   Tokyo   Toronto

Library of Congress Cataloging-in-Publication Data

TK
7874.75
.T78
1996

Tsividis, Yannis.
   Mixed analog-digital VLSI devices and technology : an introduction
/ Yannis Tsividis.
     p.   cm.
   Includes index.
   ISBN 0-07-065402-6 (alk. paper)
   1. Integrated circuits—Very large scale integration—Design and
construction. 2. Integrated circuits—Very large scale integration—
Mathematical models. I. Title.
TK7874.75.T78 1995
621.39'5—dc20

95-42375
CIP

McGraw-Hill

*A Division of The **McGraw·Hill** Companies*

Copyright © 1996 by The McGraw-Hill Companies, Inc. All rights
reserved. Printed in the United States of America. Except as permitted
under the United States Copyright Act of 1976, no part of this publica-
tion may be reproduced or distributed in any form or by any means, or
stored in a data base or retrieval system, without the prior written permission of the publisher.

1 2 3 4 5 6 7 8 9 0 DOC/DOC 9 0 1 0 9 8 7 6

ISBN 0-07-065402-6

*The sponsoring editor for this book was Steve Chapman, the editing
supervisor was Jane Palmieri, and the production supervisor was
Suzanne W. B. Rapcavage. It was set in Century Schoolbook by North
Market Street Graphics.*

Printed and bound by R. R. Donnelley & Sons Company.

The following material from the book by Y. Tsividis, *Operation and Modeling of the MOS Transistor,* McGraw-Hill, New York, 1987, has been
reprinted by permission from the publisher: From Chap. 4, Sec. 4.13 and
Figs. 1, 3, 11, 12, 15, 17, 18, 23, 25, 27, 29, 30, 31; from Chap. 5, Fig. 5;
from Chap. 8, Figs. 1, 3, 9, 16, 18, 24, 25.

Information contained in this work has been obtained by The
McGraw-Hill Companies, Inc. ("McGraw-Hill") from sources be-
lieved to be reliable. However, neither McGraw-Hill nor its authors
guarantees the accuracy or completeness of any information published herein and neither McGraw-Hill nor its authors shall be
responsible for any errors, omissions, or damages arising out of use
of this information. This work is published with the understanding
that McGraw-Hill and its authors are supplying information but are
not attempting to render engineering or other professional services.
If such services are required, the assistance of an appropriate professional should be sought.

This book is printed on acid-free paper.

Contents

Preface

Mixed analog-digital chips are now commonplace. They can be found in a variety of applications, from TV sets and compact disk players to automobiles, telephony equipment, and the disk drive of personal computers. The successful design of such chips requires, in addition to knowledge of circuits and systems, a background in device models, fabrication technology, and layout, with the special considerations that apply to analog and mixed analog-digital circuits. This book has been written to provide this background. It aims at providing just the information missing in this respect from digital VLSI textbooks and, in large part, even from traditional analog IC design textbooks and courses. The book emphasizes intuition and practical information that is found to be useful in the development of correctly working chips.

Although the book does not exhaust the subject, it does contain enough information for a good start. After reading the book, one should be able to answer questions that keep coming up in design settings in universities and companies—for example, why simulations disagree with hand calculations and with experimental results; where common simulator models, appropriate for digital circuits, fail for analog circuits, why and what to do about it; what are the limits of strong inversion in MOS transistors, and what happens outside such limits; how to model devices at nanoampere levels for low-power design; what are the frequency limits of validity of common device models found in simulators; what devices other than transistors are provided in common fabrication processes, and what the performance of such devices is; how to lay out transistors, resistors, and capacitors for good matching; how far two devices can be on the chip before matching between them deteriorates; and how to prevent interference from the digital circuits to the analog circuits on the same chip.

The book can find use in several different settings. It can be used as a supplement to courses on analog circuit design (usually senior/graduate level, using, for example, texts by Gray and Meyer, Gregorian and Temes, Allen and Holberg, or Laker and Sansen), with the material in the book covered in class or assigned as reading. This book can also be used as a main text in VLSI laboratory classes, helping lift the student's spirit above the dryness caused by layout editor manuals and design rule compilations. Finally, the book is appropriate for self-study by practicing engineers and university researchers who want to expand their activities into the mixed analog-digital VLSI field. The book has been written mainly for those who have taken a course on digital VLSI design or who have an equivalent background. However, enough introductory material has been included to make the book readable even by motivated readers whose only knowledge of electronics is that provided by basic courses. A solutions manual will be available from the publisher.

It is a feature of most prefaces to list the reasons why the subject of the book is important to study, to justify the choice of topics, and to go through the material chapter by chapter. Given that readers often bypass book prefaces, the author has decided to include such material in Chapter 1, to increase the chances of it being read.

Acknowledgments

The material in this book has been tried successfully in various university and industrial courses, starting with an early version in 1990 at Columbia University. I wish to thank all those who read parts of it and gave me useful comments, notably Dimitri Antoniadis of MIT; John Khoury, Ken Suyama, and Charles Zukowski of Columbia University; Franz Dielacher and Bert Seger of Siemens; and Marcel Pelgrom of Philips. Technical discussions with George Hart of Columbia University, George Katopis of IBM, Harry Lee of MIT, Dimitri Paraskevopoulos of National Semiconductors, David Scott of Texas Instruments, and Vance Archer, Venu Gopinathan, and Hongmo Wang of AT&T are also gratefully acknowledged. I want to thank Franz Dielacher, Venu Gopinathan, and John Khoury for providing chip photographs. Finally, thanks are due to the Alkinoos Group and to Ilya Yusim for their expert typesetting and to Tom Judson, Elsa Sanchez, and George Efthivoulidis for doing the artwork.

Yannis P. Tsividis

Mixed Analog-Digital VLSI Devices and Technology

Introduction: Mixed Analog-Digital Chips*

1.1 The Role and Place of Modern Mixed Analog-Digital Chips

This is a book about devices and technology for mixed analog-digital chips. We will not use the common term *mixed-signal* to characterize such chips, in order to avoid the connotation of ambiguity that this term has in figurative speech; also, this term does not make explicit what kinds of signals are mixed and is often used in a sense that is not broad enough. Thus, we will stick to *mixed analog-digital,* and, for brevity, we will be talking of MAD chips. Such chips can be made by using very large-scale integration (VLSI) technologies, and they can contain over 1 million transistors and other devices.

In the early 1970s, it was already clear that the pervasiveness of the digital computer had begun. To some, this observation meant that the role of analog circuits would diminish, and in fact the decline of analog circuits was forecast in technical journals, trade magazines, and even the popular press. Such predictions never materialized. What happened instead was that the pervasiveness of the digital computer became a key factor for the *increased* pervasiveness of analog, and eventually MAD, circuits. There are two main reasons for this, and both are discussed below.

1. *The need to interface the computer to the analog world.* Digital computers do not operate in a vacuum; as they pervade our physical,

---

* This chapter is based on Ref. 1.

analog world, they must interface with it. In most instances where a new application for the computer is found, sophisticated analog circuits are needed for at least the interface; thus analog pervasiveness expands. In the automobile, for example, analog electronics used to be limited to the radio; today it is found in addition in systems used for computerized engine control, safety, etc., and it is beginning to appear in new applications such as navigation aids. The computer also made possible cellular telephony; thus analog circuits, which are essential in the transmission and reception of both analog and digital signals (see below), have taken to the streets along with the telephones. In many such instances, system designers attempt to place the required analog circuits on the same chip with the digital circuits. Where feasible, this marriage can offer several advantages, as will be explained in Sec. 1.2. We should emphasize that interfacing to the analog world involves not only analog-to-digital (A/D) and digital-to-analog (D/A) converters, but also pre- and postconversion signal conditioning (amplification, filtering for antialiasing or smoothing, sampling, holding, multiplexing-demultiplexing, etc.), as well as direct signal processing without conversion to digital, especially in combinations of medium signal-to-noise ratio with high speed and/or low power [2].

The signals that must be processed by the analog part of MAD chips come from a variety of sensors, receiving antennas, transmission lines, other circuits, etc. They must drive a variety of activators, transmitting antennas, transmission lines, or other circuits. The variety of both sources and destinations for analog signals is expanding due to extensive work on sensors and actuators [3]. The expanding role of sensors in modern systems (e.g., automotive) suggests an expanding role of MAD circuits in such systems.

The advent of computers has driven the development of analog circuits not only in terms of new applications, but also in terms of performance for existing applications. Thus, for example, in the early days of digital processors, if the signal to be processed was analog, one might have been content with a modest-performance A/D converter; but when better digital processors were developed, the need arose for much better A/D converters, in terms of both resolution and speed, in order to take advantage of the increased performance offered by digital technology. The development of such converters was then successfully undertaken, entailing both the improvement of the performance of existing schemes and the invention of new ones.

Most MAD chips have been realized in metal-oxide semiconductor (MOS) technology, due to the advantages that this technology offers for digital circuits. Initially the idea of MAD MOS circuits sounded pretty mad, as there existed a prejudice that MOS transistors on chips were only good for switches. Eventually, MAD circuits took off, following quick and wide industrial acceptance of certain developments in the mid-1970s. The main such developments [4–12] were the precision-ratio capacitor array, the internally compensated MOS operational amplifier, and the use of such circuits in novel A/D converters, pulse-code modulation encoders/decoders (PCM codecs), and switched-capacitor filters. The first truly massively produced and used MAD MOS chips were PCM codecs for telephony, which were offered by several manufacturers starting in the late 1970s.

2. *The need for analog-enhanced digital performance.* In addition to helping the computer interface with the analog world, analog circuits have another role that also helped their pervasiveness: They help make possible the high performance of the digital systems and circuits themselves. A good example of an analog-enhanced system is the personal computer. Let us look at an important component of it: the hard disk drive. Here individual high-speed digital pulses, distorted due to storage and retrieval imperfections, are reshaped by analog means since digital processing would make necessary much higher-speed digital circuits, which are not feasible [13]. Within the read channel of a disk drive one may need continuous-time filters, automatic gain control circuits, phase-locked loops, etc. In recent disk drives, these are all placed on one MAD chip. Another application where MAD chips are used for the signal processing of digital pulses is in digital communication links [14].

The influence and importance of analog circuits can also be seen in what are considered as purely digital chips. Designers of high-performance digital chips will attest to the fact that high-speed digital circuits are partly analog in nature. "Square" digital waveforms taking only the values of 0 and 1 may exist in the pages of textbooks, but not in real, high-speed digital chips, in which the rise and fall times can be a large part of the total period. To handle the resulting continuous waveforms and push up clock speeds, analog considerations are needed. Thus, analog circuits are used in today's digital VLSI chips to implement efficient buffers, to modify an apparent line impedance, to aid in clock recovery, to sense and amplify minute voltage differences on memory chips, and so on. The reader may want to consult detailed descriptions of memory chips [15] and even microprocessor chips [16], in which one encounters entire analog blocks such as amplifiers, phase-locked loops, and charge pumps. We thus have the curious phenomenon that as traditional analog application areas are invaded by digital system solutions, what makes the latter possible with high performance is circuit design that is partly analog.

Our definition of MAD chips in this book includes the chips discussed in both items 1 and 2 above, although it is not common to include the latter in such a definition.

1.2   Advantages of Mixing Analog and Digital Circuits on the Same Chip

At this point it may be clear why analog circuits are needed, along with digital, in a single system. But, at least in the case of circuits handling purely analog signals (see item 1 in Sec. 1.1), a question arises: Why put such circuits on the same chip with digital circuits? Why not put analog circuits on one chip, digital circuits on another, and then interconnect the two? The answer is that, in certain cases, this two-chip solution will in fact be preferable (for example, in order to reduce the interference from the digital to the analog circuits); however, where feasible, a single-chip solution offers several advantages:

1. *The size of the system is reduced.*   This is important when electronics must fit in limited space, such as in a handheld phone or a hearing aid.

2. *The speed of operation can be increased.*   The analog and digital circuits do not have to communicate via the pins of separate packages and the wiring between them, which are plagued by parasitic capacitances.

3. *The power dissipation is decreased.*   One does not need power-hungry drivers for driving the capacitances of the pins and external wiring.*

4. *The design flexibility is increased.*   Since the analog and digital circuits are on the same chip, there is practically no limit to the number of communication channels between them.

5. *The reliability is increased.*   Fewer packages and external connections can mean a reduced probability of failure.

6. *The system cost is reduced.*   The use of mixed analog-digital chips means fewer packages, simpler printed-circuit boards, and easier assembly.

Some MAD chips contain large analog parts. In others, almost the entire chip is digital, with only a small portion devoted to analog circuits. This small portion, though, can provide added value that can make a real difference.

1.3   Applications of MAD Chips

MAD chips are used today in practically every broad application area in which chips are used. Such areas include:

---

* Another possibility is to employ a multichip module in which both chips are placed and interconnected in a single package. Such an approach will reduce the first three problems, but not as much as a single-chip approach will.

Telecommunications: Switching, transmission and reception (including fiber optics and wireless communications), end-user terminals, etc.

Consumer electronics: Audio and video equipment, appliances.*

Computers and related equipment: Local-area network transmitters and receivers, modems, disk drives, wireless communications.

Multimedia: Here the three areas just considered—namely, telecommunications, consumer electronics, and computers—meet.

Automotive systems: Engine control, safety equipment, displays, entertainment, etc.

Biomedical instrumentation: Cardiac pacemakers, hearing aids, prosthetic devices, etc.

Robotics: Interfaces to sensors and actuators.

The role of MAD chips in some of the above applications is changing. For example, there is a trend toward digital video processing in TV receivers, but critical MAD chips are being developed for the radio-frequency circuits, A/D conversion, preconversion filtering, and display driving. When "all-digital" high-definition television (HDTV) becomes a reality, MAD circuits will be needed in the input stages, in display drivers, and, of course, in high-performance "digital" chips (see item 2 in Sec. 1.1). As analog circuits move partially out of some areas, such as television, they invade other areas, such as the high-volume area of disk drive electronics, which we already mentioned above, and portable wireless communications.

Examples of MAD chips are shown in Figs. 1.1 to 1.3 [17–19]. Many other examples of MAD chips can be found in Refs. 20 to 26.

Research and development on MAD circuits, in the context of the above applications, is currently underway in low-voltage and low-power operation, very high-speed circuits, A/D and D/A conversion, continuous-time filters, current-mode circuits, noise and interference reduction techniques, self-correction techniques, etc. Research is also underway on the use of MAD circuits in emerging areas of information processing and control, such as fuzzy logic [27], neural networks [28, 29], chaotic circuits [30], and combinations thereof [31].

---

* The categories mentioned even include areas which the public (and the advertising hype) considers purely digital. Thus, for example, in a compact disk player, MAD circuits may be found in the pulse preamplifier, the D/A converter, the random access memory (RAM), the servo controller, and the power driver.

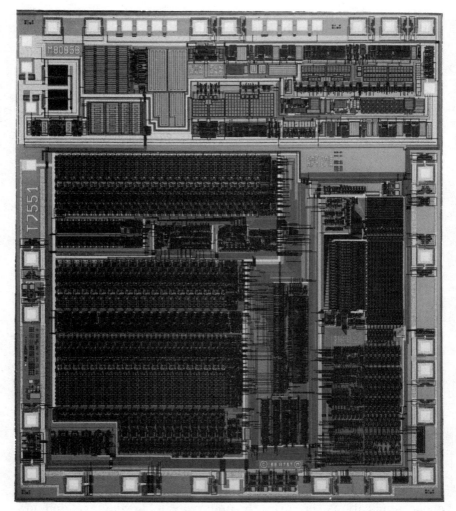

Figure 1.1 A mixed analog-digital chip [17] used for pulse-code modulation digital telephony. The chip, implemented at AT&T Bell Laboratories, includes compression/ expansion and echo cancellation functions. The analog part, seen at the top of the chip, includes sigma-delta A/D and D/A converters, voltage references, filters, and a line driver *(Copyright 1990, by IEEE.)*

1.4 Obstacles in the Design of MAD Chips

Despite the broad spectrum of applications in which MAD chips are used, such chips have not yet realized their full potential. Several reasons for this are now examined.

1. The first reason is related to the fact that analog circuits rely on the details of device characteristics to a much greater extent than dig-

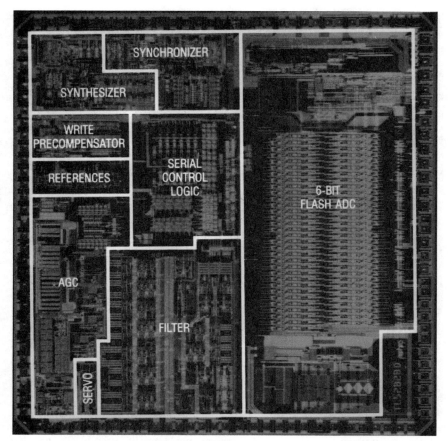

Figure 1.2 A "mostly analog" chip [18] containing all the analog front-end functions of a 21- to 64-Mbit/s hard disk drive channel plus some digital support circuitry. The chip, implemented at Texas Instruments, includes an automatic gain control (AGC) block, a 6- to 33-MHz programmable filter, two 24- to 72-MHz phase-locked loops, a differential 6-bit A/D converter with a 24- to 72-MHz sampling rate, and a write precompensator. *(Copyright 1994, by IEEE.)*

ital circuits do. The design of analog circuits generally requires very careful device modeling, which often is just not available. In contrast, in digital circuits the details of device characteristics do not play as big a role, since transistors are being switched between the on and off states and an approximate model can be used in the transition between the two.*

---

* The reliance of analog circuits on the details of device characteristics is viewed here as a problem, but it can be turned into an asset when one uses certain device characteristics to advantage; e.g., exponential characteristics of transistors can be used directly to achieve the mathematical operation of exponentiation, or taking the logarithm.

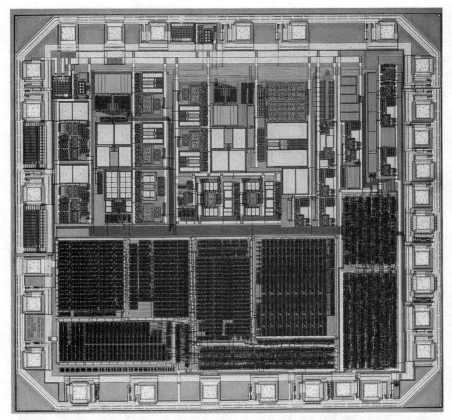

Figure 1.3 A mixed analog-digital chip containing several functions needed for a cordless telephone (Siemens). The chip includes programmable amplification blocks, encoder/ decoder, decimator, and interpolator, and it can interface directly with a microphone and a loudspeaker. The analog part is in the upper half of the chip. *(Courtesy of F. Dielacher)*

2. On MAD chips, analog circuits which are often called on to process very small signals (sometimes in the microvolt range) must be placed near digital circuits, which can produce severe interference (sometimes in the volt range). If one is not careful, the digital circuits will prevent the analog circuits from functioning properly, and thus the MAD chip as a whole will malfunction, preventing the digital circuits on it from realizing their full potential. In that case, MAD might as well stand for "mutually assured destruction." Even though, over the years, several techniques have been devised to reduce on-chip interference, the problem is still severe. This problem is compounded by the fact that its modeling is very difficult, as interference can come through a variety of parasitic paths which are not well characterized. This limits the size and complexity of MAD chips in some cases and deters some designers from even attempting to undertake the development of MAD chips.

3. By its nature, analog circuit design has many degrees of freedom and is not amenable to standardization. Thus, for example, whereas there are a few good ways to design a digital counter, there is a practically endless variety of operational amplifier designs, with each having something to offer. To a lesser extent, this situation can also be observed at the systems level. The same problem, namely, the large number of degrees of freedom, is responsible for an endless variety of impedance levels, signal swings, etc., at the input and output of analog cells; thus, interface standardization is largely absent in analog circuits. This renders the design of MAD chips a complicated affair and makes transfers from one designer to another difficult (such as when a designer leaves a company in the middle of a project).

4. For analog and MAD circuits, design automation is nowhere near where it is for digital circuits. Among the factors contributing to this problem are the high degree of required accuracy, the difficulty to precisely model device characteristics and interference, as well as the endless degrees of design freedom and the lack of standardization. The difficulties encountered are both in automating known design techniques and in coding new design knowledge into computer-aided design programs. This has its good side: It is not likely that MAD circuit designers will be replaced by design tools in the foreseeable future.

5. In contrast to digital circuits, the performance of analog and MAD circuits cannot be predicted with certainty by only testing them with a manageable number of input waveforms. Thus, testing is a very difficult problem with analog and MAD chips, which in general grows with chip size and complexity.

6. The problem to be discussed next is nontechnical and has to do with prejudices against analog circuits in the minds of many people. Analog is still viewed as it used to be in the early 1970s, that is, as inaccurate, prone to drifts, etc. Design techniques such as reliance on precision ratios and self-correcting schemes have successfully dealt with such problems, but this fact still fails to be appreciated by nonexperts. Thus, engineers and often entire companies are just not aware of what analog and mixed analog-digital designs can do for them.

7. MAD chip design expertise is not easy to come by. It is obvious from what has already been mentioned that the idiosyncracies of MAD chip design are many. To successfully deal with them, a significant amount of experience is required. In addition, due to the prejudices discussed in item 6, most universities do not provide adequate education in MAD circuits, and most students choose not to specialize in them.

The difficulties involved in designing MAD chips can often be traced to inadequate understanding of device models, technology, and layout techniques as they apply to analog and MAD circuits. When it comes to device modeling, this problem cannot be compensated for by computer simulation, since even the device models used in several simulators are inadequate and sometimes downright wrong. This is especially true of certain models for the MOS transistor, which were developed with mainly the needs of digital designers in mind. Sound model development for analog work must take explicitly into account the peculiarities of analog circuits. The same is true for layout techniques. For example, in digital circuits, the matching of device characteristics is usually not a concern; in most analog circuits, though, special layout techniques must be used to attain as high a degree of matching as possible.

1.5   The Aim and Contents of This Book

As already discussed, a successful designer of MAD chips needs to know circuit and system design [24–26, 32–47] and have an adequate, complementary background in device modeling, technology, and layout techniques as they specifically apply to such chips. This book aims at supplying this background.

The first topic undertaken in the following chapters is device modeling. Rather than cover several devices with equal (and necessarily limited) weight, we have chosen one device—the MOS transistor—and used it as a vehicle to illustrate adequately the many facets of modeling and model use. The MOS transistor is the transistor available in complementary MOS (CMOS) technology, the workhorse of VLSI; it also happens to be the transistor for which, for reasons already discussed, the models found in simulators and circuits texts are poorest. In our treatment we provide intuition about this device, we present models that go beyond those found in texts on VLSI and even analog circuit design, and we discuss the many pitfalls and misuses of models found in popular simulators. (We do not, however, describe particular simulator models in detail; books devoted to such descriptions are already widely available [48].) The material in this part of the book is influenced by the trend toward low-voltage, low-power battery-operated equipment. Thus, we pay attention to not only the traditional region of operation (strong inversion), but also the regions of weak and moderate inversion, which are more suitable for circuits used in such equipment. In addition, having in mind the importance of chips for wireless communications, we discuss the high-frequency limitations of popular models. To manage such coverage in reasonable space, we forgo the model equation derivations in most cases, since deriving such relations as we present requires a book of its own [49].*

---

* References are, of course, provided for those interested in the derivation of the results presented. The reader may already have encountered some derivations in VLSI and semiconductor device courses, at least for simple results.

Even though we omit derivations, the coverage of MOS transistors takes three chapters. Chapter 2 introduces the MOS transistor and provides a qualitative view of it, with emphasis on intuition; Chap. 3 deals with direct-current (dc) models; and Chap. 4 is devoted to small-signal modeling, including noise. We have chosen to present most equations in table form for easy reference. Several "fine details" about models, which may not be appreciated at first sight, but the knowledge of which can prove very valuable when particular problems are faced during design, are given in footnotes, both in the text and the tables.

We hope that by the end of Chap. 4, readers will have at their disposal models that work well for analog and digital designs and will have acquired the "feel" and attitude needed to judge other models encountered in the course of their work. After studying this material, readers should have no problem dealing with models of other devices, such as bipolar junction transistors or integrated resistors and capacitors, which are summarized later in the book in less detail.

The next topic is fabrication technology. Chapter 5 is devoted to it. Here we concentrate exclusively on silicon technologies, which are the standard in the industry and are the ones used in MAD chips for most applications. CMOS is singled out for emphasis since it is the most widely used technology, thanks to its excellent combination of high performance and low cost. We go through the essential fabrication steps and discuss possible enhancements. We then consider the performance and modeling of various devices other than MOS transistors, which are available to the designer in this technology in the context of MAD circuits, and we discuss various considerations, such as tolerances, that are important in that context. We also consider other silicon fabrication technologies, concentrating mostly on the bipolar-CMOS (BiCMOS) process, which makes available both bipolar and MOS transistors. Since this process can be viewed, in part, as an extension of CMOS technology, our treatment of it is short. After showing typical BiCMOS processes, we discuss the modeling of the bipolar junction transistor. This device is better covered than the MOS transistor in many electrical engineering curricula (with adequate information on it even provided in circuit design texts [38]), and it is better modeled in simulators. In addition, parts of Chaps. 3 and 4 are general and apply to the bipolar junction transistor, or to any other transistor for that matter; these include, e.g., general discussions of the process of modeling, discussions on the meaning of small-signal parameters, the general introduction to noise, and discussions of parameter extraction and its pitfalls. Thus the treatment of the bipolar junction transistor in Chap. 5 is brief. For some readers previously exposed to this device, this treatment will be a convenient review.

The final chapter, Chap. 6, deals with layout. To make this material readable by those with no previous experience in this topic, we introduce it from ground zero; most of the chapter, though, is spent on the aspects of layout that are particular to analog and MAD circuit design. This includes our discussion of the factors affecting matching and of ways to achieve it, as well as the discussions of digital-to-analog interference and of the ways to prevent it. Readers can learn more about layout, with emphasis on digital circuits, from sources on digital VLSI [32–37]; many readers probably have already been exposed to such sources.

Analog and digital circuit design is described in many well-known texts [32–47]. Starting from such knowledge and complementing it with the material in this book, readers should be able to undertake the design of successful mixed analog-digital chips. We wish them every success in this endeavor.

References

1. Y. P. Tsividis, "R & D in Analog Circuits: Possibilities and Needed Support," *Proceedings of European Solid-State Circuits Conference '92*, Copenhagen, September 1992, pp. 1–5.

2. E. A. Vittoz, "Low-Power Design: Ways to Approach the Limits," *Digest of Technical Papers, 1994 IEEE International Solid-State Circuits Conference*, San Francisco, February 1994, pp. 14–18.

3. K. T. V. Grattan, *Sensors and Actuators—Technology, Systems, and Applications*, Adam Hilger, Bristol, England, 1992.

4. J. L. McCreary and P. R. Gray, "All MOS Charge Redistribution Analog-to-Digital Conversion Techniques—Part I," *IEEE Journal of Solid-State Circuits*, 10: 371–379, December 1975.

5. Y. Tsividis, P. R. Gray, D. A. Hodges, and J. Chacko, "An All-MOS Companded PCM Voice Encoder," *Digest of Technical Papers, IEEE International Solid-State Circuits Conference*, Philadelphia, PA, February 1976, pp. 24–25.

6. F. H. Musa and R. C. Huntington, "A CMOS Monolithic 3 Digit A/D Converter," *Digest of Technical Papers, IEEE International Solid-State Circuits Conference*, Philadelphia, PA, February 1976, pp. 144–155.

7. Y. Tsividis and P. R. Gray, "An Integrated NMOS Operational Amplifier with Internal Compensation," *IEEE Journal of Solid-State Circuits*, 11:748–754, December 1976.

8. J. A. Young, D. A. Hodges, and P. R. Gray, "Analog NMOS Sampled-Data Recursive Filter," *Digest of Technical Papers, IEEE International Solid-State Circuits Conference*, February 1977, pp. 156–157.

9. J. T. Caves, M. A. Copeland, C. F. Rahim, and S. D. Rosebaum, "Sample Analog Filtering Using Switched Capacitors as Resistor Equivalents," *IEEE Journal of Solid-State Circuits*, 12:592–599, December 1977.

10. B. J. Hosticka, R. W. Brodersen, and P. R. Gray, "MOS Sampled Data Recursive Filters Using Switched Capacitor Intergrators," *IEEE Journal of Solid-State Circuits*, 12:600–608, December 1977.

11. D. A. Hodges, P. R. Gray, and R. W. Brodersen, "Potential of MOS Technologies for Analog Integrated Circuits," *IEEE Journal of Solid-State Circuits*, 13:285–294, June 1978.

12. Y. P. Tsividis, "Design Considerations in Single-Channel MOS Analog Integrated Circuits—A Tutorial," *IEEE Journal of Solid-State Circuits*, 13:383–391, June 1978.

13. Session devoted to disk drive electronics, *Digest of Technical Papers, 1994 IEEE International Solid-State Circuits Conference*, San Francisco, February 1994, pp. 276–287.

14. T. H. Hu and P. R. Gray, "A Monolithic 480 Mb/s Parallel AGC/Decision/Clock-Recovery Circuit in 1.2 μm CMOS," *IEEE Journal of Solid-State Circuits*, 28:1314–1320, December 1993.

15. K. Ihibashi, K. Tagasugi, T. Yamanaka, T. Hashimoto, and K. Sasaki, "A 1-V TFT-Load SRAM Using a Two-Step Word-Voltage Method," *IEEE Journal of Solid-State Circuits*, 27:1519–1524, November 1992.

16. I. A. Young, J. K. Greason, and K. L. Wong, "A PLL Clock Generator with 5 to 110 MHz of Lock Range for Microprocessors," *IEEE Journal of Solid-State Circuits*, 27: 1599–1607, November 1992.

17. V. Friedman, J. M. Khoury, M. Theobold, and V. P. Gopal, "The Implementation of Digital Echo Cancellation in Codecs," *IEEE Journal of Solid-State Circuits*, 25:979–986, August 1990.

18. D. Choi, R. Pierson, F. Trafton, B. Sheahan, V. Gopinathan, G. Mayfield, I. Ranmuthu, S. Venkatraman, V. Pawar, O. Lee, W. Giolma, W. Krenik, W. Abbott, and K. Johnson, "An Analog Front-End Signal Processor for a 64 Mbits/s PRML Hard-Disk Drive Channel," *IEEE Journal of Solid-State Circuits*, 29:1596–1605, December 1994.

19. F. Dielacher, private communication.

20. *IEEE Journal of Solid-State Circuits*, various issues; especially the December issue of each year, which emphasizes analog circuits.

21. *Digest of Technical Papers, IEEE International Solid-State Circuits Conference*, various years.

22. P. R. Gray, D. A. Hodges, and R. W. Brodersen, eds., *Analog MOS Integrated Circuits*, IEEE Press, New York, 1980.

23. P. R. Gray, B. A. Wooley, and R. W. Brodersen, *Analog MOS Integrated Circuits*, vol. 2, IEEE Press, New York, 1988.

24. J. Trontelj, L. Trontelj, and G. Shenton, *Analog Digital ASIC Design*, McGraw-Hill, London, 1989.

25. R. S. Soin, F. Maloberti, and J. Franca, eds., *Analog-Digital ASICs*, Peter Peregrinus, London, 1993.

26. J. E. Franca and Y. Tsividis, eds., *Design of Analog-Digital VLSI Circuits for Telecommunications and Signal Processing*, Prentice-Hall, Englewood Cliffs, NJ, 1994.

27. J. W. Fattaruso, S. S. Mahant-Shetti, and J. B. Barton, "A Fuzzy Logic Inference Processor," *IEEE Journal of Solid-State Circuits*, 29:397–402, April 1994.

28. Y. Arima, M. Murasaki, T. Yamada, A. Maeda, and H. Shinohara, "A Refreshable Analog VLSI Neural Network Chip with 400 Neurons and 40K Synapses," *IEEE Journal of Solid-State Circuits*, 27:1854–1861, December 1992.

29. C. Mead, *Analog VLSI and Neural Systems*, Addison-Wesley, Reading, MA, 1989.

30. A. Rodriguez-Vasquez, "CMOS Design of Chaotic Oscillators Using State Variables: A Monolithic Chua's Circuit," *IEEE Transactions on Circuits and Systems*, pt. 2, 40:596–613, October 1993.

31. Y. Horio and K. Suyama, "IC Implementation of Switched-Capacitor Chaotic Neuron," *Proceedings of the International Symposium on Circuits and Systems, London*, 1994, pp. 97–100.

32. N. Weste and K. Eshragian, *Principles of CMOS VLSI Design*, Addison-Wesley, Reading, MA, 1993.

33. L. A. Glasser and D. W. Dobberpuhl, *The Design and Analysis of VLSI Circuits*, Addison-Wesley, Reading, MA, 1985.

34. J. P. Uyemura, *Fundamentals of MOS Digital Integrated Circuits*, Addison-Wesley, Reading, MA, 1988.

35. M. Shoji, *CMOS Digital Circuit Technology*, Prentice-Hall, Englewood Cliffs, NJ, 1988.

36. D. A. Hodges and H. G. Jackson, *Analysis and Design of Digital Integrated Circuits*, McGraw-Hill, New York, 1988.

37. E. D. Fabricius, *Introduction to VLSI Design*, McGraw-Hill, New York, 1990.

38. P. R. Gray and R. G. Meyer, *Analysis and Design of Analog Integrated Circuits*, Wiley, New York, 1993.

39. R. Gregorian and G. C. Temes, *Analog MOS Integrated Circuits for Signal Processing*, Wiley-Interscience, New York, 1986.

40. P. E. Allen and D. G. Holberg, *CMOS Analog Circuit Design,* Holt, Rinehart and Winston, New York, 1987.
41. J. Davidse, *Analog Electronic Circuit Design,* Prentice-Hall International, New York, 1991.
42. A. B. Grebene, *Bipolar and MOS Analog Integrated Circuit Design,* Wiley-Interscience, New York, 1984.
43. M. R. Haskard and I. C. May, *Analog VLSI Design,* Prentice-Hall, New York, 1988.
44. R. Unbenhauen and A. Cichocki, *MOS Switched-Capacitor and Continuous-Time Integrated Circuits and Systems,* Springer-Verlag, New York, 1989.
45. R. L. Geiger, P. E. Allen, and N. R. Strader, *VLSI Design Techniques for Analog and Digital Circuits,* McGraw-Hill, New York, 1990.
46. C. Toumazou, F. J. Lidgey, and D. G. Haigh, eds., *Analogue IC Design: The Current-Mode Approach,* Peter Peregrinus, London, 1990.
47. K. R. Laker and W. M. C. Sansen, *Design of Analog Integrated Circuits and Systems,* McGraw-Hill, New York, 1994.
48. P. Antognetti and G. Massobrio, *Semiconductor Device Modeling with Spice,* McGraw-Hill, New York, 1993.
49. Y. Tsividis, *Operation and Modeling of the MOS Transistor,* McGraw-Hill, New York, 1987.

The MOSFET: Introduction and Qualitative View

2.1 Introduction

In this chapter we describe a simplified structure of a MOS transistor, and we offer a qualitative, intuitive description of how this device operates. This material will give the reader a bird's-eye view of MOS transistor behavior and will naturally lead to the qualitative models of Chap. 3 or even to other qualitative models that the reader is likely to encounter in the course of circuit design activities.

In many circuits, MOS transistors, when turned on, are operated with rather large currents in the *strong inversion* region of operation. In fact, this region is the only one to which attention is paid in most circuit design texts. The MOS transistor can, however, also be operated with minute currents, in a region of operation called *weak inversion.* This region is very important for low-power, battery-operated equipment and for today's low-supply-voltage technologies. Thus we make sure to cover it adequately. The same holds for the *moderate inversion* region between the weak and strong inversion regions.

2.2 MOS Transistor Structure

A simplified structure of an *n-channel* MOS transistor is shown in Fig. 2.1 (the names *n-channel* and *MOS* will be discussed shortly). The transistor is formed on a silicon *body* or *substrate* in which dopant (impurity) atoms have been introduced.* The dopant atoms used here

---

* In our discussion, basic semiconductor topics such as dopants, free electrons, and holes are explained only in passing. More details can be found elsewhere [1–5].

are *acceptors,* called so because each accepts one valence electron from a silicon atom, leaving in the latter an electron vacancy, called a *hole.* The hole (being the absence of a negatively charged electron) behaves as a positively charged particle which, at common operating temperatures, is free to wander around. Due to the positive charge associated with each hole, the acceptor dopants and the substrate material as a whole are referred to as *p type.* Since the dopant atoms "release" holes into the substrate by accepting valence electrons from the environment, each dopant atom acquires a net negative charge. In contrast to the hole, though, this negative charge is *bound* to the atom; i.e., it is immobilized and is *not* free to wander around. A typical doping concentration for the body is 10^{16} dopant atoms per cubic centimeter (often written simply as 10^{16} cm^{-3}).* The dopant concentration will be assumed uniform throughout the body, until further notice.

The two regions indicated as *source* and *drain* in Fig. 2.1 (typically a few tenths of a micrometer deep) are doped with *donor* impurity atoms. At common operating temperatures, each of these atoms releases an electron to its environment, and thus the donor atoms and the regions doped with them are called *n-type* regions. Each donor atom that releases an electron is left with a net positive charge. That positive charge is associated with the immobile donor atom and thus is a "bound" charge which cannot move, in contrast to the electron released by that atom. The source and drain regions are heavily doped, which is indicated with the notation n^+ in the figure. The heavy doping results in low resistivity for these regions, since an abundance of electrons are available for conduction.

---

* Exponents in units apply to the units as a whole. Thus cm^{-3} means (cm)$^{-3}$, μm^2 means $(\mu m)^2$, etc.

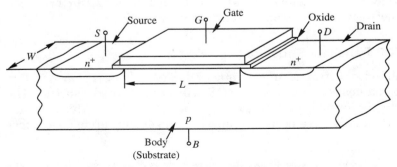

Figure 2.1 Simplified structure of an *n*-channel MOS transistor.

The region between the source and drain in Fig. 2.1 is called the *channel*. In an analog circuit, the channel *width* W and *length* L of individual transistors can range from a fraction of a micrometer to over 1000 μm, depending on the fabrication technology and circuit design needs. Although channel lengths below 0.1 μm have been demonstrated in the laboratory, such lengths are not yet used in production. The channel is covered by an insulating layer, typically silicon dioxide (and commonly referred to simply as *oxide*). The insulator thickness in modern fabrication processes is typically 70 to 200 Å (0.007 to 0.02 μm). The body interface to the oxide is often called the *surface*.

A low-resistivity electrode, called the *gate*, is formed on top of the oxide. Contemporary processes commonly use polycrystalline silicon (*polysilicon*, or *poly*, for short) for the gate. In such processes (referred to as *silicon gate processes*), the polysilicon gate is, in fact, formed *before* the source and drain regions. The source and drain regions are formed by implanting donor atoms, with the gate acting as a "mask" against the implant; this "mask" receives the donor atoms itself and prevents them from landing under it. Thus the gate is doped n^+ and exhibits low resistivity. The donor atoms land just outside the "shadow" of the gate. Subsequent high-temperature fabrication steps, though, cause a diffusion of the dopant atoms both vertically and laterally. The *lateral diffusion* causes the source and drain regions to extend slightly under the gate, as shown in Fig. 2.1. The resulting *overlap* distance is a few tenths of a micrometer.

As we will see below, if the gate potential is made sufficiently positive with respect to other parts of the structure, electrons can be attracted directly below the insulator (near the "surface" of the body). These electrons can come through the n^+ regions, where electrons exist in abundance, and can fill the channel between them; for this reason the device in Fig. 2.1 is referred to as an *n-channel* device. (The opposite type of device, called *p-channel*, has holes in its channel and will be considered later.) The number of electrons in the channel can be varied through the gate potential. This can cause a variation of the "strength" of the connection between the two n^+ regions, resulting in "transistor action." If the two n^+ regions are biased at different potentials, the lower-potential n^+ region acts as a source for electrons, which then flow through the channel and are drained by the higher-potential n^+ region. It is thus common to call the lower-potential n^+ region *source* and the higher-potential one *drain*, although not everyone abides by this convention in all instances.

The source, gate, drain, and body regions can be contacted through terminals attached to them, as shown schematically in Fig. 2.1. We denote these terminals by S, G, D, and B, respectively.

The first successful MOS transistor used metal for the gate material and silicon dioxide for the insulator. It is for this reason that the device was named *MOS transistor,* with MOS standing for *metal-oxide semiconductor.* Other acronyms are *MOST* (for MOS transistor), *MOSFET* (for MOS field-effect transistor), and *IGFET* (insulated gate field-effect transistor). The latter name originated to distinguish the device from the junction gate field-effect transistor, in which the gate is separated from the rest of the structure by a *pn* junction. Of the three acronyms, IGFET is the most general, since it does not specify the material used for the gate or the insulator. This acronym, though, is not in wide use among integrated circuit (IC) designers. Today, the popular acronyms MOST and MOSFET have come to mean the same as IGFET, and they do not imply that metal and silicon dioxide are used for the gate and the insulator.

Often *n*-channel and *p*-channel MOS transistors are referred to simply as *nMOS* and *pMOS* transistors, respectively. These short names will be favored in this book.

2.3   Assumptions about Terminal Voltages, Currents, and Temperature

In integrated circuits fabricated by using common processes, many MOS transistors are formed on a common substrate. These transistors must be isolated from each other (unless they are intentionally meant to be connected). The substrate serves as a common isolating region. However, that region forms *pn* junctions with the source and drain regions; if these junctions are forward-biased (i.e., if the *n* regions become negative in potential with respect to the *p* regions), large currents will flow, interfering with the intended isolation and with transistor action. For this reason one must ensure that the junctions never become forward-biased. For an *n*MOS transistor (Fig. 2.1), this requires the following conditions:

$$V_{SB} \geq 0 \qquad\qquad (2.1)$$

$$V_{DB} \geq 0 \qquad\qquad (2.2)$$

where V_{SB} and V_{DB} denote the source-body and drain-body voltages, respectively. Until further notice, all voltages and currents are assumed to be "dc" (constant). Such dc quantities are denoted by capital variables with capital subscripts.

The above conditions are implicitly assumed for all *n*MOS transistors in our discussions, unless stated otherwise. These conditions guarantee that only a very small reverse-bias *leakage* current flows in the *pn* junctions (typically less than 1 pA for minimum-size regions at room temperature; note, though, that this current increases with temperature, roughly doubling with every 8 to 10°C temperature increase). The current flowing through the gate insulator is even smaller than the above current, in fact by orders of magnitude. Thus unless stated otherwise, we assume that leakage currents are negligible.

In the discussion that follows, the two n^+ regions are allowed to be at possibly different potentials. We follow the common convention mentioned in the previous section, and we call the *drain* the n^+ region which is at the higher potential of the two. We thus assume that $V_{DB} \geq V_{SB}$, allowing for the special case of equal potentials. Since $V_{DB} - V_{SB} = V_{DS}$, this is equivalent to

$$V_{DS} \geq 0 \qquad (2.3)$$

The above assumption can be made without loss of generality. If the polarity of the voltage between the two regions is reversed, the names of the two regions can also be reversed and the statements made below about *source* and *drain* remain valid.

One last assumption that holds for all the discussions in this book is that all parts of a transistor operate at the same temperature. Unless otherwise mentioned, this is assumed to be *room temperature,* taken as 300 K (degrees Kelvin). Transistor behavior at other temperatures is considered in Chap. 3.

2.4   A Qualitative Description of MOSFET Operation

Over the years, many different models have been developed for describing MOS transistor operation quantitatively. A circuit designer is likely to be faced with several of these (e.g., simple models for hand analysis and complicated ones for computer simulation, and different ones for different regions of operation). What is more, models change over the years as new ones are developed or installed in computer simulators. For this reason, it is counterproductive to tie our initial description of transistor operation to a particular model. What we do instead is to provide first a general, qualitative description, which is more or less applicable to any model. Quantitative modeling is considered in Chap. 3.

2.4.1   Effect of V_{GS}: Level of inversion

Let us consider a transistor connected to external voltage sources as in Fig. 2.2a. The drain is short-circuited to the source, making $V_{DS} = 0$; the more general case of $V_{DS} > 0$ is discussed later. The external source-body bias V_{SB} is fixed at a positive value V_{SB1} and, since the source is short-circuited to the drain, acts as the reverse bias for both *pn* junctions. Let us concentrate on one of the two junctions, say, the one associated with the source. The shadow around this junction indicates the presence of a *depletion region,* i.e., a region from which most "mobile" carriers have been removed, with holes in the body having moved away from the junction toward the negative terminal of the V_{SB} battery and electrons in the n^+ regions having moved in the opposite direction toward the positive terminal of the battery. This means that, on the *p* side, immobile acceptor atoms are left with a net negative charge (they have been "uncovered"); also, on the n^+ side, immobile donor atoms are left with a net positive charge. The total charges revealed on each side of the junction are equal in magnitude, preserving charge neutrality. Because the n^+ region is much more heavily doped, only a very narrow strip need be uncovered on that side, to balance the charges on the much deeper strip

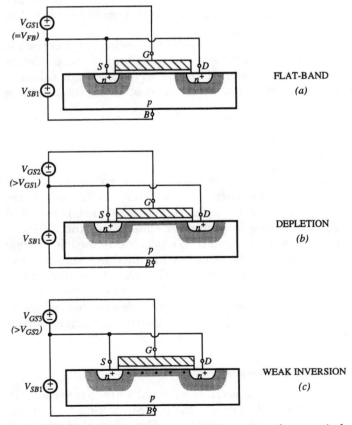

Figure 2.2 The MOS transistor with $V_{DS} = 0$ and successively increased values of V_{GS} (indicated by increased numerical subscripts).

on the p side. In the figure, only the depletion region on the p side is shown, and it is indicated by the shadow around the n^+ region.

The depth of the depletion region is related to the charge density in it and the electrostatic potential across it, according to basic electrostatic laws [1–5]. This depth is such that the resulting electrostatic potential drop across the region, along with the *contact potentials* at the contacts to the device, and the external bias V_{SB} algebraically add up to zero. A depletion region will exist even with $V_{SB} = 0$; the potential across it balances the contact potentials around the loop. If V_{SB} is increased, the depletion region will deepen accordingly to support the extra drop. It can be shown that since the n^+ region is much more heavily doped than the p region, practically all the potential drop across the depletion region occurs on the p side.

On the drain side, another depletion region forms in the same way. Since the drain is short-circuited to the source, the drain-body junction sees the same reverse bias as the source-body junction, and thus both depletion regions have the same depth.

If V_{GS} is sufficiently negative, positively charged holes can be attracted at the surface (the substrate interface to the oxide); this condition is known as *accumulation*. If the surface is accumulated, and subsequently V_{GS} is made less and less negative, the accumulation will become lighter and lighter, and at a certain value of V_{GS} it will disappear altogether, leaving the body neutral. This is the *flat-band* condi-

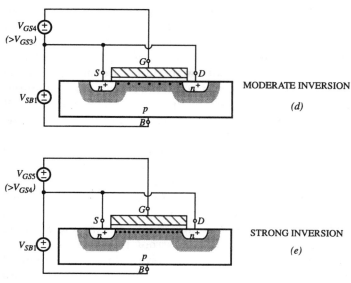

Figure 2.2 *(Continued)*

tion, assumed in Fig. 2.2a. The corresponding value of V_{GS} is called the *flat-band voltage* and is denoted by V_{FB}. At the flat-band condition, each immobile negatively charged acceptor atom in the body outside the two depletion regions is, on average, covered by a mobile, positively charged hole, resulting in macroscopic neutrality.

If V_{GS} is now made more positive than its flat-band value, the charge on the gate will become more positive, too. This will tend to repel holes from the surface, resulting in a depletion region in the channel as shown in Fig. 2.2b. This condition is called *depletion*.

If V_{GS} is raised still farther, more positive charges will be placed on the gate, which must be balanced by more negative charges in the channel; to uncover the extra negative charges, the depletion region in the channel must become deeper. Assume now that V_{GS} is increased to the point that the depletion region in the channel becomes almost as deep as that under the n^+ regions. This means that the electrostatic potential at the surface (with respect to the body) is now almost the same as that on the n^+ regions. Thus the surface becomes almost as attractive for electrons as the n^+ regions. Electrons now enter from the n^+ regions into the channel, next to the surface, despite the fact that the body is p type. The surface is now said to be *inverted,* and the electrons next to the surface form what is known as an *inversion layer.* The larger the value of V_{GS}, the more the electrons are attracted to the surface and the "heavier" the inversion.

Although the inversion level increases continuously as V_{GS} is raised, the range of inversion levels (and the corresponding ranges of V_{GS} values) is divided into three regions, in each of which the transistor will later be seen to exhibit distinct properties. The three regions are called *weak, moderate,* and *strong* inversion and are illustrated in Fig. 2.2c, d, and e, respectively. The solid dots in the channel represent electrons. The number of dots, however, is not meant to be proportional to the number of electrons; no drawing could accomplish that, since the number of electrons can vary by orders of magnitude from one level of inversion to the next. The distinction between the three levels of inversion will remain qualitative for now; it will become quantitative later.

Note that as V_{GS} is raised, the increased charges on the gate must be balanced by more negative charges under the insulator. In weak and moderate inversion, the available electrons are not enough to accomplish this task by themselves, and thus the depletion region deepens significantly, uncovering more acceptor atoms with negative charge in order to contribute to the balance. This is evident, for example, in going from c to d or from d to e in Fig. 2.2. Once strong inversion is reached, though, the electrons find it extremely easy to enter the channel through the source and drain regions. If V_{GS} is raised still further into strong inversion, practically all the new gate charges are balanced by more electrons entering the inversion layer. The depletion region depth now changes only slightly, and thus so does the potential vertically across it; these quantities are said to be practically *pinned* to a fixed value in strong inversion. In fact, in strong inversion the electrons are in such abundance that the inversion layer behaves somewhat as an n^+ region. For this reason, the name *field-induced n^+p junction* is sometimes used for the combination of the strong inversion layer and the body. In strong inversion, the gate-oxide-inversion layer combination behaves in a manner similar to that of a two-plate capacitor.

2.4.2 Effect of V_{SB}: The body effect

Let us now consider the effect of V_{SB}. Recall that (1) the value of V_{SB} determined how deep the depletion region would be under the n^+ regions and (2) the depletion region under the channel had to be made almost as deep (by increasing V_{GS}) before significant numbers of electrons would flow into the channel. If, in Fig. 2.2a, the value of V_{SB} is raised, the two depletion regions under the n^+ regions will become deeper. To reach a similar depletion region depth under the channel, a larger V_{GS} value than before will be needed. Electrons will start flowing to a significant extent into the channel only if that larger V_{GS} value (and the corresponding larger depletion region depth in the channel) is reached; weak, moderate, and strong inversion will each be attained at larger V_{GS} values than before.

In the case of *strong* inversion, there is another way to see this. Consider the situation in Fig. 2.3a, where a strong inversion layer is assumed to have been formed. Here V_{SB} acts as a reverse bias for the "field-induced junction" formed by the strong inversion layer and the body. Thus if V_{SB} is raised, the depletion region of that "junction" will widen, uncovering more acceptor atoms and revealing their negative charge. Now fewer electrons will be needed in the inversion layer to balance the gate charges, as illustrated in Fig. 2.3b. If it is desired to restore the number of electrons to what it was before V_{SB} was raised, then V_{GS} will have to be increased, as shown in Fig. 2.3c.

The above effect of the channel-body bias on the inversion level is called the *body effect,* or *substrate effect.* Since $V_{BS} = -V_{SB}$, increasing V_{SB} means making the body potential more negative with respect to the source. Thus decreasing the body potential (with respect to the source) decreases the number of electrons, just as decreasing the gate potential (again, with respect to the source) would. The body thus acts somewhat as a gate, and for this reason it is sometimes called the *back gate.* The body effect is sometimes referred to as the *back-gate effect.*

If the body doping concentration is larger, the number of extra acceptor atoms that are uncovered (with increasing V_{SB}) can be large, and

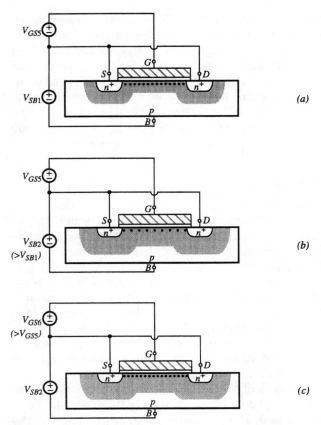

Figure 2.3 Illustration of the *body effect* in strong inversion: (*a*) original bias, assumed the same as in Fig. 2.2*e;* (*b*) effect of an increase in V_{SB}; (*c*) effect of a subsequent increase in V_{GS}, aimed at restoring the original level of inversion.

the body effect can be severe. This can be compensated for if the oxide is sufficiently thin, since then the gate is close to the channel and can keep better control of the inversion layer charge (as opposed to releasing this control to the "back gate"). These qualitative claims are verified by quantitative analysis [6].

It will be seen later that, due to the body effect, an increase in V_{SB} results in a deterioration of the channel conduction, other things being equal.

2.4.3 Effect of V_{DS}: Drain current

We now insert a voltage source V_{DS} between the source and the drain, with $V_{DS} > 0$. We assume that this is done in each of the cases shown in

Fig. 2.2, without altering the other voltages. We then have the situations shown in Fig. 2.4, where the same V_{DS} value (denoted by V_{DS1}) is assumed in all cases. The details shown in this figure will now be explained.

First, note that in all cases the depletion region under the source is still the same as in Fig. 2.2, since the reverse bias there (V_{SB}) is assumed to be the same. Since corresponding parts of Figs. 2.2 and 2.4 are assumed to have the same V_{GS} (and the same V_{SB}) values, the inversion level *near the source* is the same in them, being weak in Fig. 2.4c, moderate in Fig. 2.4d, and strong in Fig. 2.4e. It is customary to use these names to describe the situations in these figures without explicit reference to the source. Thus, for example, Fig. 2.4e is said to illustrate "a MOSFET in strong inversion" although, as we will see, if V_{DS} is large, strong inversion is guaranteed only near the source. We follow this convention.

To explain what happens on the drain side, let us start from Fig. 2.4a. We have two depletion regions, as in Fig. 2.2a. However, the reverse bias of the drain-body junction is now $V_{DB} = V_{DS} + V_{SB}$, which is larger than V_{SB} since V_{DS} is positive. Thus the depletion region under the drain is wider than that under the source, as shown. Raising V_{GS} begins to deplete the channel, too, as shown in Fig. 2.4b.

In Fig. 2.2c (weak inversion), surface electrons appeared near the source in nonnegligible numbers because V_{GS} was large enough to make the depletion region in the channel almost as deep as that under the source. The same thing happens in Fig. 2.4c, *near the source*. However, the channel depletion region is *not* as deep as the *drain* depletion region, which means that near the right-hand end of the channel the situation is not so favorable for inversion.* Recall now that the depletion region under the drain is deeper because the drain is at a more positive potential (with respect to the body) than points in the channel. Thus any electron that finds itself in the vicinity of the drain is attracted quickly to the drain owing to this higher potential. Electrons then enter the channel from the source, diffuse through the channel, and are drained quickly by the drain, being replenished by new electrons entering from the source. A current is now seen to flow in the channel and the external battery V_{DS}. The movement of electrons in the channel is due to *diffusion,* which takes place because there is a *gradient* of concentration along the channel, starting from a significant concentration near the source and reducing to near-zero concentration near the drain. This diffusion is similar to the diffusion of smoke particles from a region of high concentration to one of low concentration,

---

* This is a manifestation of the body effect discussed in the previous subsection; however, the term *body effect* is often reserved for the effect of V_{SB} only.

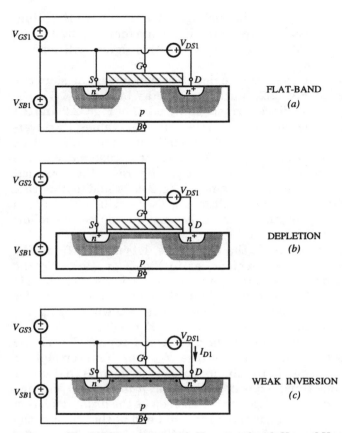

Figure 2.4   The MOS transistor with $V_{DS} > 0$ and with V_{SB} and V_{GS} as in the corresponding parts of Fig. 2.2. Increased numerical subscripts in V_{GS} indicate increased values.

and it does not need a horizontal electric field in the channel in order to occur. Near the source, where the concentration is significant, the electrons move slowly; near the drain, their concentration is lower, but they move faster. Hence, the same current can be supported at all points in the channel. One can think of smoke in a pipe; lightly concentrated smoke moving quickly past a point can carry the same number of smoke particles per unit time as heavily concentrated smoke that is moving more slowly.

The movement of the negatively charged electrons occurs from the source, through the channel, to the drain. This is equivalent to a conventional *positive* current in the opposite direction. Thus the *drain current* I_D, defined as shown in Fig. 2.4c, is positive (assuming always that V_{DS1} is positive).

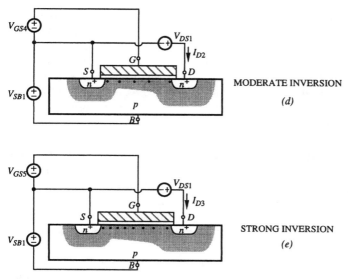

Figure 2.4   *(Continued)*

If V_{GS} is raised further, we enter moderate inversion (strictly speaking, the source end of the channel does so), as shown in Fig. 2.4*d*. Now more electrons are available for current flow, resulting in a larger drain current. The electrons continue to move to the right due to diffusion, but *not only* due to diffusion. Now the inversion layer begins, in addition, to have properties reminiscent of those of resistive materials. It partly acts as a resistor, in which current can be conducted owing to the electric field caused by the potential drop across it*—this type of conduction is referred to as *drift*. The current in moderate inversion is thus due to *both* diffusion *and* drift.

Finally, if we make V_{GS} large enough, we enter strong inversion, as illustrated in Fig. 2.4*e*. The abundance of electrons here makes possible an even larger drain current. Now the inversion layer can really behave as a (nonlinear) resistor, resulting in a large drift current, in comparison to which the diffusion contribution is insignificant. In strong inversion, then, the current flows mostly by *drift*. The value of this current can be orders of magnitude larger than in weak inversion. The "reverse bias" across the "field-induced n^+p junction" formed by the inversion layer and the body is approximately equal to V_{SB}

---

* This "resistor" is nonlinear, a fact that will be appreciated when we discuss the dependence of I_D on V_{DS}.

near the source and gradually increases to $V_{DB} = V_{DS} + V_{SB}$ near the drain. Thus the depletion region depth under the inversion layer increases accordingly from left to right, as shown. This means that near the drain a larger amount of immobile, uncovered, negatively charged acceptor atoms are to be found. Fewer electrons are thus needed near the drain to balance the gate charges, as shown. In fact, depending on the value of V_{DS}, the density of electrons near the drain can be so low that the channel there is only moderately or weakly inverted, although the transistor is still said by convention to be "in strong inversion" because the *source* end of the channel is strongly inverted. The electrons move progressively faster as they approach the drain, and thus the same amount of current can be carried at any position in the channel. At sufficiently large values of V_{DS}, the electrons lose a significant amount of energy due to collisions, and their velocity can reach a limit; this phenomenon is referred to as *velocity saturation.*

Consider now Fig. 2.4*c, d,* and *e.* If V_{DS} starts from a very small value and is increased in any one of them, the current increases. In weak inversion, this is because the higher drain potential reduces the electron concentration near the drain even further, thus increasing the magnitude of the concentration gradient along the channel and the resulting diffusion current. In strong inversion, the larger voltage drop across the inversion layer causes larger fields in it and a larger drift current. In moderate inversion, both effects come into play. As V_{DS} is increased from zero upward, a plot of I_D versus V_{DS} looks qualitatively as indicated by the initial portion of the curve in Fig. 2.5. This qualitative behavior is observed in any region of inversion, although the quantitative details differ from region to region. Figure 2.5 is intended only to indicate the general shape of the curve.

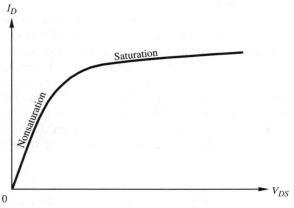

Figure 2.5 I_D versus V_{DS}, for fixed V_{GS} and V_{SB}.

Saturation. For large enough V_{DS} values, the current gradually tends to saturate, as indicated by the part of the curve to the right in Fig. 2.5. To see the reason for this, recall that given a large enough drain potential, the electron concentration in the inversion layer near the drain becomes very small. The channel at that end is often said to be "pinched off" in this case, although, of course, the electron concentration there never becomes exactly zero. It just becomes very small, and the necessary current is supported by an appropriately large electron velocity. Further increases in V_{DS} cannot affect things much, because the electron concentration is already negligible near the drain and remains negligible as V_{DS} is increased. Thus the rest of the inversion layer "does not notice" much difference toward the drain end, and the current does not change much, as shown to the right in Fig. 2.5.

Two regions can thus be distinguished, marked *nonsaturation* and *saturation* in the figure. Although these two terms are often used only in the context of strong inversion, we use them in any region of inversion in this book.* Although the transition between nonsaturation and saturation is smooth, their boundary is sometimes taken to be at a specific point for convenience (Chap. 3).

The exact mechanisms of current conduction in saturation are complicated, but often the following simplified picture is used. Let us assume that the channel has become practically "pinched off" when V_{DS} reaches a certain value denoted by V'_{DS}, as shown in Fig. 2.6a. If now V_{DS} is increased above V'_{DS}, the inversion layer charge near the drain decreases even further. The pinched-off tip of the channel thus moves to the left, as shown in Fig. 2.6b, and the region between that end and the drain becomes approximately a depletion region. Note that this region is not completely depleted; some fast-moving electrons are still found in it at any given time (although in the oversimplified model reported here, the region is commonly shown as in Fig. 2.6b). The inversion layer to the left of that region cannot support more voltage than V'_{DS}, since it becomes pinched off when the voltage across it reaches that value. The excess voltage $V_{DS} - V'_{DS}$ is thus dropped between the drain and the tip of the channel, as shown in Fig. 2.6b. If V_{DS} is increased further, then the voltage drop across the depletion region must increase and the depletion region expands to support it. Thus the inversion layer to the left of that region effectively shrinks somewhat in length. This is referred to as *channel length modulation*. Since now the voltage V'_{DS} is dropped over a shorter distance, horizon-

---

* Other terms sometimes used for these two regions in the strong inversion case are *triode region* and *pentode region,* from the names of electronic vacuum tubes that exhibit characteristics qualitatively similar to those in nonsaturation and saturation, respectively.

tal electric field intensities in the channel increase and so does the drift current. Also, since the electron concentration in the channel must now change from its source-end value to its pinched-off value over a shorter distance, the magnitude of the electron concentration gradient increases, which increases the diffusion current. Thus in all regions of inversion the drain current keeps increasing somewhat with V_{DS}, *even* in the saturation region. This effect has been included in Fig. 2.5.

If the actual channel length L is large, then the shrinkage of the inversion layer is a negligible fraction of L, and thus the device "does not really notice" much of a difference as V_{DS} is increased; the inversion layer remains practically the same length (in a relative sense), with practically the same voltage across it (V'_{DS}). Thus the current in saturation increases by only a minute amount. In short-channel devices, though, the shrinkage of the inversion layer is a significant fraction of L and the current increases considerably. Also, these effects are more severe when the body doping N_A is small, since then the length of the pinched-off region is larger (the scarcer ionized dopant atoms must be uncovered over a longer distance, so that the excess voltage $V_{DS} - V'_{DS}$ can be supported in accordance to basic electrostatics).*

---

* As already mentioned, the pinched-off idea illustrated in Fig. 2.6 is actually an over-simplification. To accurately look at what happens in the vicinity of the drain, one must take into account velocity saturation and must consider the electric field distribution in detail. As it turns out from computer two-dimensional solutions, in this region the electrons dip below the surface and reach the drain at various depths.

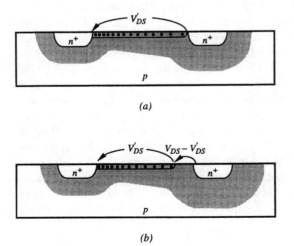

(a)

(b)

Figure 2.6 Part of the MOS transistor in saturation, illustrating the common but oversimplifying "pinched-off" idea: (a) $V_{DS} = V'_{DS}$; (b) $V_{DS} > V'_{DS}$.

2.5 A Fluid Dynamical Analog

To help increase intuition about the MOS transistor, we now present a fluid dynamical analog [7, 8]. Let us begin with strong inversion. In our analog, electrons correspond to molecules of a fluid. Electric current corresponds to the net flow of that fluid. The source and drain correspond to two large tanks, filled with this fluid up to a certain level each. As fluid is moved from one tank to the other, the level of fluid in each tank is kept constant through external means. This corresponds to the potentials at the source and drain of a transistor being held constant despite current flow. We assume that the level of fluid in the "source tank" is fixed throughout our discussion. The two tanks are separated by a piston corresponding to the gate. This is shown in Fig. 2.7a. A "handle," shown by a centerline, is attached to the piston. The depth of this handle with respect to the "source" level is denoted by $\hat{V}_{GS}$ and corresponds to the gate-source voltage. The lower the handle (and thus the piston), the larger the value of $\hat{V}_{GS}$. Similarly, the level of fluid in the "drain" tank with respect to the source level is denoted by $\hat{V}_{DS}$ and corresponds to the drain-source voltage. The lower the "drain" level, the larger the value of $\hat{V}_{DS}$.

In Fig. 2.7a, $\hat{V}_{GS}$ is very low. The "channel" is cut off, and no communication exists between source and drain. Now assume that $\hat{V}_{GS}$ is increased, as shown in Fig. 2.7b. The channel is now filled with the fluid. Communication between the source and drain is now possible, but no flow is observed in the steady state, since $\hat{V}_{DS} = 0$. Now if $\hat{V}_{DS}$ is increased as in Fig. 2.7c, then flow is observed as shown. The flow increases as $\hat{V}_{DS}$ is increased further, until "saturation" is reached; in saturation (Fig. 2.7d), further increases in $\hat{V}_{DS}$ do not affect the flow. Note that the fluid enters slowly from the source and, as it approaches the drain, moves faster in order to maintain a fixed flow, despite the fact there is less fluid near the drain. Again one can make an analogy to the simplified picture of transistor operation in saturation.

It is evident from the figure that, for a given $\hat{V}_{DS}$, the flow will increase if $\hat{V}_{GS}$ is increased and will decrease if $\hat{V}_{GS}$ is decreased. If $\hat{V}_{GS}$ is decreased such that the top surface of the piston is at or slightly above the source surface level, then direct flow of the fluid is prevented. Water molecules, though, can still flow from left to right. This can be seen if one takes into account the diffusion of water vapors [8], as illustrated in Fig. 2.8a. This situation corresponds to weak inversion. As shown, the vapor concentration is maximum at the water surface and decreases as one moves vertically away from it (in fact, it can be shown that it decreases exponentially). Thus, if the water surface in the drain tank is lower, as shown in Fig. 2.8a, at any given horizontal plane above the piston, the vapor concentration will be decreasing as one goes from left to right over the channel. Vapor thus

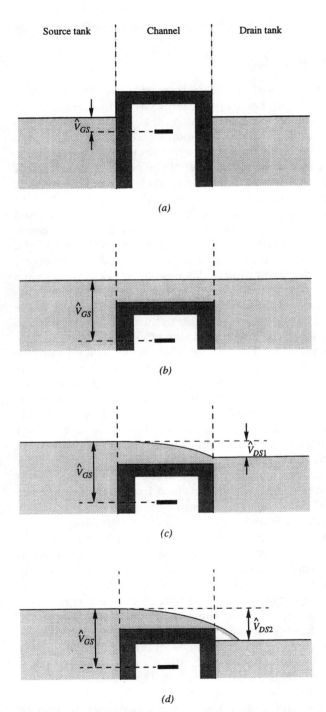

Figure 2.7 Fluid dynamical analogs of MOS transistor operation in cutoff and strong inversion: (a) cutoff; (b) with channel strongly inverted, but no current since $\hat{V}_{DS} = 0$; (c) nonsaturation with nonzero $\hat{V}_{DS}$; (d) saturation.

diffuses from left to right, carrying a minute "current," even if the position of the middle shaft is slightly above the source water surface. However, a very small drop in the position of the middle shaft (i.e., an increase in $\hat{V}_{GS}$) can drastically increase the "current," since the vapor concentration over the source, at points slightly above the level of the shaft, will increase exponentially (Fig. 2.8b). One thus might expect an exponential dependence of the "current" on shaft position and, if the analogy is valid, an exponential dependence of I_D on V_{GS} in weak inversion in the real transistor. This expectation will indeed be born out shortly. For large $\hat{V}_{DS}$ values in Fig. 2.8 (low drain water level), concentration on the right at the level of the shaft's surface becomes negligible, and the "current" assumes a value largely independent of $\hat{V}_{DS}$, reaching saturation.

If the top surface of the piston is only *very* slightly *below* the source level, both the fluid and its vapors can contribute significantly to the

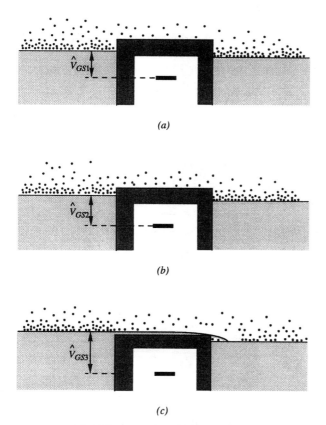

(a)

(b)

(c)

Figure 2.8   A fluid dynamical analog of MOS transistor operation in weak inversion (a and b) and moderate inversion (c).

flow (Fig. 2.8c). This corresponds to moderate inversion, in which both drift and diffusion contribute to the current.

2.6 Complete Set of Characteristics

Let us finally look at a complete set of typical transistor characteristics. In Fig. 2.9a we show plots of I_D (on a *logarithmic* axis) versus V_{DS}, with V_{GS} as a parameter, and $V_{SB} = 0$, for a specific transistor. The logarithmic axis is used to reveal the several orders of magnitude of I_D over which control is possible through V_{GS}. The regions of inversion have been roughly marked in terms of V_{GS}, for later use (the limits between regions will be discussed soon). As V_{GS} is reduced, the current in the channel can eventually become so small that it is masked by the leakage current of the reverse-biased drain-body junction (or even the leakage from the inversion layer to the substrate). This can be seen near the bottom of the plot. The boundary between the regions of nonsaturation and saturation is indicated by a broken line; saturation is to the right of that line. Note that in weak inversion the onset of saturation occurs at the same V_{DS} value, independent of the value of V_{GS}; this is not true for moderate and strong inversion.

The same characteristics are shown in Fig. 2.9b where a *linear* I_D axis is used. Obviously this type of plot does not do justice to weak

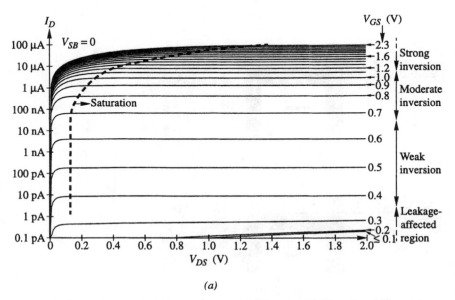

(a)

Figure 2.9 The I_D-V_{DS} characteristics for a specific device, with $V_{SB} = 0$ and V_{GS} a parameter: (a) logarithmic I_D axis.

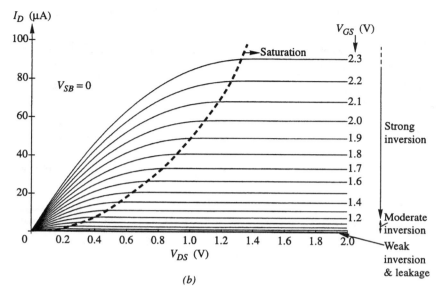

Figure 2.9 (*Continued*) The I_D-V_{DS} characteristics for a specific device, with $V_{SB} = 0$ and V_{GS} a parameter: (*b*) linear I_D axis.

inversion. This region was, in fact, unknown for several years, until suspicious "leakage currents" in dynamic memories prompted researchers to take a closer look. The weak inversion region is rather well understood by now, and it is an important region of operation for many applications, especially those for which very small power dissipation, or low-voltage operation, is desired.

If V_{SB} is raised, the *body effect* will reduce the electron concentration in the inversion layer and the drain current, for a given V_{GS} and V_{DS}. This is illustrated in the plots of Fig. 2.10, which have been obtained for $V_{SB} = 2$ V. The body effect becomes evident when these plots are compared to the corresponding ones in Fig. 2.9.

The above plots have been obtained for a long-channel device. For this device, the increase of I_D with V_{DS} in saturation (due to the channel length modulation effect discussed in Sec. 2.4.3) is not easily seen on the scale used. As already explained, channel length modulation becomes more evident for shorter channels; an example is shown in Fig. 2.11. Until further notice, though, we will assume that in the devices we consider this effect is negligible.

2.7 Form of Functional I_D-V_{GS} Dependence: Practical Limits for Regions of Inversion

A careful look at transistor characteristic curves reveals a distinctly different behavior in each region of inversion. Let us begin with the

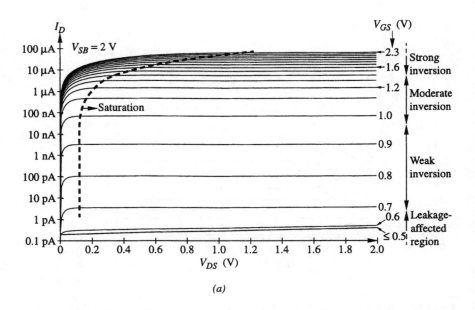

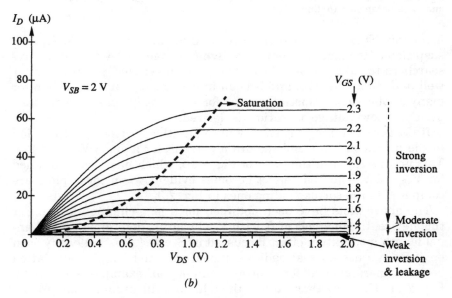

Figure 2.10 The characteristics of the device used for Fig. 2.9, with V_{SB} changed to 2 V: (a) logarithmic I_D axis; (b) linear I_D axis.

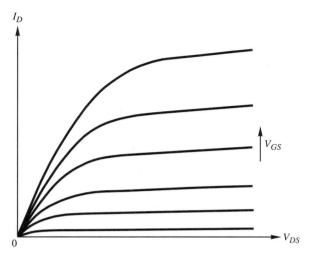

Figure 2.11 Typical I_D-V_{DS} characteristics (with V_{GS} a parameter) for a device with evident channel length modulation.

weak inversion region, which is marked on Figs. 2.9a and 2.10a. The spacing of consecutive curves in this region can be seen to be almost equal, for equal V_{GS} increments. Given that the vertical axis is logarithmic, I_D is, for all practical purposes, exponentially related to V_{GS} in weak inversion. This behavior reminds one of the type of dependence of collector current on base-emitter voltage in a bipolar transistor.

To make the above functional dependence even more evident, let us fix the value of V_{SB} and go up the curves vertically at a fixed V_{DS}, as illustrated in Fig. 2.12a by the broken line at $V_{DS} = V_{DS1}$. We can pick the values on each curve and plot I_D (again on a logarithmic axis) versus V_{GS}, as shown in Fig. 2.12b. The straight-line part of this plot indicates exponential I_D-V_{GS} behavior. Below this part, exponentiality is destroyed because of the interference of leakage currents. Above the straight-line part, exponentiality is destroyed because drift currents begin to become important (i.e., we enter the moderate inversion region).

In this book, we will indicate only approximate limits for each region of inversion. For weak inversion, the upper limit will be taken at a point just below the region where the behavior begins to deviate significantly from exponentiality. Such a point is seen to be (V_M, I_M) in Fig. 2.12b. A lower limit (V_K, I_K) is also indicated as a point below which leakage currents interfere with exponentiality; the location of this point depends on the temperature and drain potential.

Typically, the width of the weak inversion region can be a few tenths of a volt; nevertheless, due to the exponential behavior in this region,

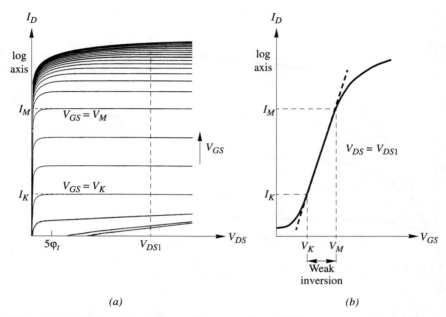

Figure 2.12 (a) I_D (on a logarithmic axis) versus V_{DS}, with V_{GS} as a parameter and fixed V_{SB}; (b) I_D (on a logarithmic axis) versus V_{GS}, as obtained from (a) with $V_{DS} = V_{DS1}$.

this is sufficient for a drain current variation of three or four orders of magnitude at room temperature.

Bypassing for the moment the moderate inversion region, we now discuss strong inversion. We concentrate only on the saturation region for now. In Fig. 2.13a we show I_D (on a linear axis) versus V_{DS} for a given V_{SB}, with V_{GS} as a parameter. We go up vertically, as indicated by the broken line at $V_{DS} = V_{DS1}$, choosing V_{DS1} so that all curves are in their saturated part. We plot $I_D - V_{GS}$ values obtained in this way in Fig. 2.13b. Finally, to reveal the functional dependence of I_D on V_{GS}, we also plot $\sqrt{I_D}$ versus V_{GS}, as shown by the solid curve in Fig. 2.13c. The curve approaches a straight line* which, when extrapolated as shown by the broken line, defines a quantity V_T noted in the figure. This quantity is known as the *extrapolated threshold voltage*. We refer to it as the *threshold voltage* for simplicity. The reader is warned that other, somewhat incompatible definitions of "threshold voltage" are also used in the literature. Those are discussed in Sec. 3.12.

If the above construction is repeated for a larger V_{SB}, the body effect will reduce the current for a given V_{GS}, and thus the curve in Fig. 2.13c will shift to the right. Thus, V_T increases with V_{SB}.

---

* We are assuming idealized behavior here. It will be seen later that in real devices the straight-line behavior is not fully observed, because the "effective mobility" of the electrons in the channel depends on V_{GS} (Sec. 3.6).

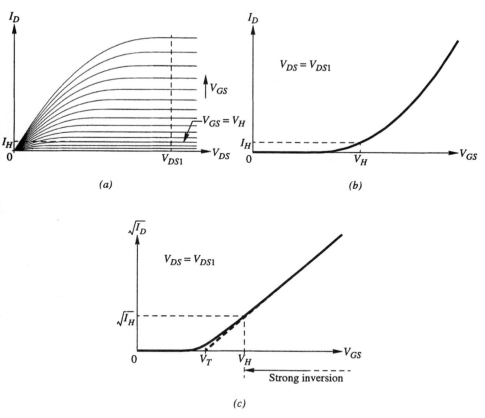

Figure 2.13 (a) I_D versus V_{DS}, with V_{GS} as a parameter: (b) I_D versus V_{GS}, obtained from part (a) with $V_{DS} = V_{DS1}$. (c) $I_D^{1/2}$ versus V_{GS}, obtained from part (b).

In the upper part of the curve in Fig. 2.13c, $\sqrt{I_D}$ is proportional to $V_{GS} - V_T$, which means that I_D is proportional to $(V_{GS} - V_T)^2$. Thus, the MOS transistor is said to be a "square-law device" in strong inversion saturation. The square-law behavior can be shown to be a consequence of the decisive presence of drift currents in strong inversion. The inversion layer charge responsible for such currents (at the bottom "plate" of the oxide "capacitor") is proportional to $V_{GS} - V_T$. The voltage across the channel (V'_{DS} in Fig. 2.6) turns out to be also proportional to $V_{GS} - V_T$ (Chap. 3). These two facts result in the current's being proportional to $(V_{GS} - V_T)^2$. The behavior deviates from the square law when the plot in Fig. 2.13c deviates from a straight line, which can be approximately taken to be the case below the point $V_{GS} = V_H$ shown there. This point can be taken as the lower limit of strong inversion. The strong inversion region can include a couple of orders of magnitude of current variation, and it extends all the way to breakdown levels (Sec. 3.9).

The values of V_H and V_M. It is obvious that the value of V_H depends on what one means above by "deviation from square law," since such deviation is completely gradual, as seen in Fig. 2.13c. The definition of V_H is kept flexible in order to conform to common practice in using strong inversion results. For example, if all one needs is a rough estimate of I_D, then a strong inversion expression (Sec. 3.2) corresponding to the broken line in Fig. 2.13c can be used down to rather low values of V_{GS}; thus V_H (the "bottom" of strong inversion) is low, too, for this application (for example, $V_T + 0.2$ V). If, however, one is designing a circuit which relies significantly on the very form of the I_D-V_{GS} dependence in strong inversion saturation, then the lower part of the curve is undesirable, and use of strong inversion equations there would be misleading and would hurt the design; thus, for this application the "bottom" of strong inversion V_H would be higher (for example, $V_T + 0.5$ V). The value of V_H should also be chosen about as high if it is viewed as a limit of validity of *all* strong inversion results collectively, including those for the small-signal parameters (Chap. 4), since some of those become inaccurate rather quickly as V_{GS} is lowered. To conclude, then, the value of V_H, viewed as the lower value of V_{GS} at which strong inversion results can be applied, depends on the application at hand and the accuracy desired.

Similar comments would seem to apply to the upper limit of weak inversion V_M; however, the uncertainty as to the value of this quantity is much smaller (for example, 50 mV) because the deviation from exponentiality in Fig. 2.12b becomes clear within a small region of V_{GS}. Usually, V_M is slightly below V_T (by, for example, 0.1 V). The values of V_T, V_M, and V_H will be influenced by fabrication process details, temperature, and the value of V_{SB}. We should note here that precise, quantitative definitions of V_M and V_H do exist [6]; they are not, however, suitable for the purposes of simple hand analysis.

What about moderate inversion? In Fig. 2.14 we repeat the plot of Figs. 2.12b and 2.13c, with the V_{GS} axes aligned for easy comparison. The moderate inversion region is shown in the figure to be between $V_{GS} = V_M$ and $V_{GS} = V_H$. In this region, the behavior is neither exponential nor square-law. This is because both drift and diffusion currents are important in moderate inversion. Over the width of the moderate inversion region, the current can vary by a couple of orders of magnitude.

Although the limits ($V_{GS} = V_M$ and $V_{GS} = V_H$) have been chosen by looking only at saturation, they are used independently of the value of V_{DS} to distinguish the regions of inversion. However, the form of functional dependence of I_D on V_{GS} in strong inversion nonsaturation is different from that in saturation. This can be seen in Fig. 2.9b; at a low, constant V_{DS} value (say, 0.2 V), equal increments of V_{GS} are seen to produce equal increments of I_D, in contrast to the square-law behavior observed in saturation.

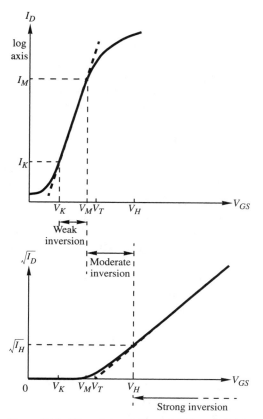

Figure 2.14 The plots of Figs. 2.12b and 2.13c with V_{GS} axes aligned. In moderate inversion, I_D is neither exponential nor square-law with V_{GS}.

2.8 Factors Affecting the Extrapolated Threshold Voltage

As indicated in the previous section, the value of V_T depends on V_{SB} due to the body effect. The special value of V_T obtained when $V_{SB} = 0$ is denoted by V_{T0} and is a quantity used widely in MOS transistor modeling and circuit design. We now discuss the fabrication process parameters that affect this value.

The quantity V_{T0} is, roughly speaking, a value of V_{GS} around which the inversion level changes from weak, through moderate, to strong. Thus V_{T0} is a measure of how easy it is to "turn on" the channel with $V_{SB} = 0$, and its dependence on various factors can be argued by similar reasoning as that used in Sec. 2.4.2 for the body effect. For example, the thinner the oxide, the smaller the value of V_{T0} (other things being equal), since the attraction that the gate can exert on electrons in the channel is stronger. Also, the lighter the body doping, the smaller the value of V_{T0}, other things being equal. To see this, consider the extreme case of zero doping (a fictitious situation); then, to balance the gate charge, all the required negative charge below the oxide would have to be provided by free electrons. These would have to enter the channel at very low values of V_{GS} since there would be no ionized acceptor atoms to share that task.

Another factor affects the value of V_{T0}: In Fig. 2.2, in series with the applied bias, we have some *contact potentials*. These are electrostatic potentials which appear when two different materials are brought into contact [1–6]. In the figure we have two relevant contacts: one to the gate and another to the body. The combined value of the corresponding contact potentials turns out to depend only on the gate and substrate materials [6]. This combined value can aid or oppose the gate bias, depending on its polarity, thus making the effective V_{T0} smaller or larger.

The value of V_{T0} is also influenced by the fact that, during fabrication, some "parasitic" positive charge is trapped at the oxide-silicon interface or even within the oxide. This charge helps the charge of the gate to attract free electrons in the channel and thus reduces V_{T0}. In contrast to this "accidental" charge, some charge can be introduced intentionally in order to adjust the value of V_{T0}, by using a process called *ion implantation* during fabrication of the device. During this process, high-energy dopant ions are accelerated toward the channel from above; they land in the channel, where they become immobile. If these ions are positive, they cancel the effect of some of the negatively ionized acceptor atoms, thus in a sense making the effective body doping concentration lighter; thus V_{T0} will be smaller. If the ions are negative instead, they help the negatively ionized acceptor atoms in balancing the gate charge, and thus fewer free electrons are needed in the balancing act; the effect is that of a heavier substrate doping, and V_{T0} becomes larger.*

By taking the above factors into account and adjusting the ion implantation accordingly, the value of V_{T0} can be made positive or negative. Common values of V_{T0} are from 0.4 to 0.8 V, but sometimes values outside this range are used, including negative ones. The *n*MOS transistors with positive V_{T0} are called *enhancement*-mode devices, since it takes a positive V_{GS} to enhance the channel with electrons so that strong inversion can be achieved. The *n*MOS transistors with negative V_{T0} are called *depletion*-mode devices, since at zero V_{GS} an inver-

---

* Ion implantation can be used not only to adjust V_{T0}, but also to "tailor" to our needs other characteristics as well, e.g., to reduce "punch-through" (Sec. 3.9).

sion layer is already formed (assuming $V_{SB} = 0$) and it takes a negative V_{GS} to deplete the channel and "turn the device off."

We remind the reader that, as explained in the previous section, the value of V_T increases above V_{T0} as V_{SB} is made positive; this is the most often mentioned manifestation of the body effect. As expected from the discussion in Sec. 2.4.2, the increase will be more pronounced for devices with large body doping and/or thick oxide.

2.9 Other Factors Affecting the Drain Current

From the discussion so far, we know that the current, for given V_{SB} and V_{DS}, will depend on how large V_{GS} is in comparison to V_T (see, for example, Fig. 2.14). Even for a given $V_{GS} - V_T$, though, the current depends on other factors. First, the amount of charge carried by electrons per unit of time depends on their velocity. For low electric fields, the latter is proportional to the electric field intensity. The constant of proportionality is the *effective mobility,* denoted by μ, which gives the velocity per unit of field intensity; it is thus measured in (cm/s)/(V/cm), or $cm^2/(V \cdot s)$. For electrons at the surface of an nMOS transistor, a typical value for μ is 600 $cm^2/(V \cdot s)$ at room temperature. The same dependence on μ is also seen in diffusion currents, because the "diffusion constant" in equations describing such currents turns out to be proportional to μ [1–5].

The current also depends on how much charge a given V_{GS} value is able to induce in the channel. This depends on the oxide capacitance per unit area, through which the charge is induced by the gate. The oxide capacitance per unit area will be denoted by C'_{ox}.

We now consider the effect of geometric dimensions on the drain current. We assume that, for the purposes of this discussion, these dimensions are not too small (we shall see in Chap. 3 what we mean by "too small," and what happens in that case).

The drain current is proportional to the channel width W. If W is doubled in Fig. 2.1, for example, we will have the equivalent of two devices, each of the original width, in parallel; thus the total current will double.

Also, the current is inversely proportional to the channel length L. For drift current, this can be seen as follows. The drift current is proportional to the velocity of electrons, which in turn is proportional to the electric field, if the latter is not too high to cause velocity saturation. The electric field along the channel results from the distribution of the total potential V_{DS} across the channel, over a distance of length L. For a given V_{DS}, the electric field at a point in the channel (at a given fractional distance from the source, say, at the middle of the channel) is inversely proportional to L (for very small V_{DS}, in fact, the field is approximately constant along the channel and equal to V_{DS}/L). Thus the drift current is also inversely proportional to L. As to the diffusion current, it is proportional to the electron concentration gradient (the rate of change of concentration with distance). That gradient is inversely proportional to L, and so is the diffusion current.*

Thus, for all levels of inversion, the drain current is proportional to the quantity W/L, which is the *aspect ratio* of the channel geometry. This ratio can be chosen practically at will, and it provides an important degree of freedom during circuit design.

This chapter was meant as an intuitive introduction to the characteristics of the MOS transistor. In principle, dc bias and low-frequency small-signal circuit design can be done directly based on measured data, such as those in Fig. 2.9 or 2.10 (a very common practice in vacuum tube days). Analytical models, though, are still desirable and prove very convenient during design and computer simulation. Also, they are a must for predictions about devices that have not yet been measured or have not even been built. The following chapter presents useful information on analytical MOS transistor models.

References

1. E. S. Yang, *Microelectronic Devices*, McGraw-Hill, New York, 1988.
2. R. S. Muller and T. I. Kamins, *Device Electronics for Integrated Circuits*, Wiley, New York, 1986.
3. B. G. Streetman, *Solid-State Electronic Devices*, Prentice-Hall, Englewood Cliffs, NJ, 1980.
4. S. M. Sze, *Semiconductor Devices: Physics and Technology*, Wiley, New York, 1985.
5. D. H. Navon, *Semiconductor Microdevices and Materials*, Holt, Rinehart and Winston, New York, 1986.
6. Y. P. Tsividis, *Operation and Modeling of the MOS Transistor*, McGraw-Hill, New York, 1987.
7. C. H. Sequin, "A Fluid Model for Visualizing MOS Transistor Behavior," sec. 1.15 in C. Mead and L. Conway, *Introduction to VLSI Systems*, Addison-Wesley, Reading, MA, 1980, pp. 29–33.
8. C. Mead, *Analog VLSI and Neural Systems*, Addison-Wesley, Reading, MA, 1989.

Problems

2.1 In Fig. 2.4e, assume that V_{SB} is changed from the value V_{SB1} to a higher value V_{SB2}. Indicate qualitatively how the sketch should be modified.

2.2 For the device used to produce the plots of Figs. 2.9 and 2.10, find graphically

---

* The fluid dynamical analog of Figs. 2.7 and 2.8 is helpful here—consider what happens if the length of the channel is increased.

(a) I_D if $V_{GS} = 0.65$ V, $V_{SB} = 0$ V, and $V_{DS} = 1$ V
(b) I_D if $V_{GS} = 1.95$ V, $V_{SB} = 2$ V, and $V_{DS} = 1.5$ V
(c) V_{DS} if $I_D = 40$ µA, $V_{SB} = 0$ V, and $V_{GS} = 1.8$ V
(d) V_{GS} if $V_{SB} = 2$ V, $I_D = 50$ µA, and $V_{DS} = 1.4$ V

2.3   For the device used to produce the plots of Figs. 2.9 and 2.10, estimate the value of the threshold voltage V_T for $V_{SB} = 0$ V and $V_{SB} = 2$ V.

2.4   In Fig. 2.9b, verify that in strong inversion saturation we have approximately square-law behavior, that is, I_D is proportional to $(V_{GS} - V_T)^2$; find the value of the constant of proportionality.

2.5   For the device used to produce the plots of Figs. 2.9 and 2.10, estimate the drain current if $V_{SB} = 2$ V, $V_{DS} = 2$ V, and $V_{GS} = 2.5$ V. Use results from Prob. 2.4.

2.6   For the data of Fig. 2.9, produce a plot like that in Fig. 2.12b. Label the axes numerically, and estimate the quantity $d(\log I_D)/dV_{GS}$ in weak inversion from this plot.

2.7   (a) For the device used to produce the plots in Figs. 2.9 and 2.10, consider the operating point $V_{SB} = 0$ V, $V_{GS} = 2.1$ V, and $V_{DS} = 0.6$ V. Estimate graphically the ratios $\Delta I_D/\Delta V_{GS}$ and $\Delta I_D/\Delta V_{DS}$ around that point, where Δ indicates small changes. (The meaning of these ratios will be discussed in detail in Chap. 4.)
(b) Repeat part (a) for $V_{SB} = 0$ V, $V_{GS} = 2.1$ V, and $V_{DS} = 1.8$ V.

2.8   From Fig. 2.9b, find $\Delta I_D/\Delta V_{GS}$ (see Prob. 2.7) around $V_{GS} = 1.55$ V and around $V_{GS} = 2.25$ V for
(a) $V_{DS} = 0.5$ V (nonsaturation)
(b) $V_{DS} = 1.8$ V (saturation)
What qualitative difference do you see between cases (a) and (b)?

3

MOSFET DC Modeling

3.1 Introduction

In Chap. 2 we described qualitatively various phenomena that are responsible for the flow of current in MOS transistors. A quantitative description of such phenomena using semiconductor physics concepts leads to quantitative models for the current in the device as a function of the externally applied bias voltages. The process of deriving such models of adequate accuracy can take a large number of pages and is the subject of other books [1–4]. Here we will present the results of such extensive derivations, and we will concentrate on their interpretation and correct use.

Our emphasis will be on simple models, appropriate for hand calculations and for fast computer simulations. We will present such simple models for each of the weak inversion and strong inversion regions. We will then show how, through interpolation, one can construct models that are valid in all regions (including moderate inversion) without excessive complexity. We will also consider a number of improvements that can be made, in order to extend the validity of the models in the presence of nonconstant mobility, temperature variations, and effects that appear when the device dimensions are made very small. We will conclude the chapter with a discussion of the present state of computer-aided design (CAD) models and of the procedures through which values are chosen for the model parameters.

We will assume that the devices considered are nMOS, unless stated otherwise. All voltages will be assumed to be dc in this chapter, and the resulting dc current will be given as a function of these voltages.* Consistent with Chap. 2, we will assume that the source-substrate and drain-substrate junctions never become forward-biased. Thus, the following conditions will be assumed to hold:

$$V_{SB} \geq 0 \qquad (3.1)$$

$$V_{DB} \geq 0 \qquad (3.2)$$

Unless stated otherwise, the devices considered will be assumed to be operating at room temperature, and their gate and substrate leakage currents will be assumed negligible. Finally, until further notice we will assume that the channel dimensions W and L are not too small (i.e., are not too close to the minimum dimensions allowed in a given fabrication process) and that the bias voltages are not too high (e.g., no more than a few volts). The need for these assumptions will become evident in Sec. 3.8.

In studying this chapter, the reader should be prepared for a considerable amount of cross-referencing and page turning. While this is inconvenient, it cannot be avoided, given our goal to expose the reader to many aspects of modeling and to the interrelations among different models, in a restricted amount of space.

3.2 DC Model for Weak and for Strong Inversion

Simple expressions that can be used in weak or strong inversion are summarized in Table 3.1 [5, 6, 1]. Using them, one can calculate the dc drain current I_D as a function of V_{GS} and V_{DS}, provided the values of several model parameters are known. Of these parameters, one (W/L) is determined by the geometry of the transistor. Of the other parameters in Table 3.1, some are dependent on the source-body bias V_{SB} due to the body effect; these are given in Table 3.2. Some of the expressions in Table 3.2 are derived in Ref. 1; others are empirical approximations of results derived in that reference. Finally, Table 3.3 gives expressions for several physical or process-dependent parameters used in the first two tables, as well as equivalent convenient formulas.†

---

* A completed dc modeling task would also involve the evaluation of charges in the transistor [1–4]. Charges are mostly employed in the evaluation of large-signal transient responses, which involves charge derivatives with respect to terminal voltages. Such derivatives are functions of the terminal voltages and, viewed as small-signal capacitances, are discussed in Chap. 4.

† The units in common use in integrated-circuit work are not consistent with common systems of units, such as SI, but have instead evolved to their present forms due to such factors as convenience and habit. Thus, e.g., channel dimensions are commonly given in micrometers (10^{-6} m), but oxide thickness is usually given in angstroms (10^{-10} m) or nanometers (10^{-9} m). Device capacitances are commonly given in femtofarads (10^{-15} F), areas in square micrometers or square millimeters, doping concentrations in cm^{-3}, etc. Due to this inconsistent use, cumbersome unit conversions are often needed. To make

TABLE 3.1   Approximate Large-Signal DC Model for Drain Current (*n*MOS Transistor)

Region	Drain current* I_D	With:
Weak inversion[†] $V_{GS} \leq V_M$ $I_{D,SAT} \leq I_M$	$I_M \exp\left(\dfrac{V_{GS} - V_M}{n\,\phi_t}\right)\left[1 - \exp\left(-\dfrac{V_{DS}}{\phi_t}\right)\right]^{\ddagger\S}$	$I_M = \dfrac{W}{L}\,I'_M$
Moderate inversion[¶] $V_M < V_{GS} < V_H$ $I_M < I_{D,SAT} < I_H$	No simple expression available; one must use general expressions (Sec. 3.5)	$I_M = \dfrac{W}{L}\,I'_M$ $I_H = \dfrac{W}{L}\,I'_H$
Strong inversion[¶¶] $V_{GS} \geq V_H$ $I_{D,SAT} \geq I_H$		$I_H = \dfrac{W}{L}\,I'_H$ $V'_{DS} = \dfrac{V_{GS} - V_T}{\alpha}$
• Nonsaturation $(V_{DS} \leq V'_{DS})$	$K\left[(V_{GS} - V_T)V_{DS} - \dfrac{1}{2}\alpha V_{DS}^2\right]$	$K = \dfrac{W}{L}\,K'$ $k = \dfrac{W}{L}\,k'$
• Saturation $(V_{DS} > V'_{DS})$	$\dfrac{k}{2}(V_{GS} - V_T)^2$ [§]	$k' = \dfrac{K'}{\alpha}$

* $V_{DS} \geq 0$ is assumed. The expressions do not include junction leakage currents and break-down effects.
[†] See Fig. 2.12.
[‡] If the accurate prediction of the dependence of I_D on V_{SB} is of interest, a different formula is preferable (Sec. 3.4).
[§] If it is desired to include the dependence of I_D on V_{DS} in saturation, use $I_D = I'_D(1 + V_{DS}/V_A)$, where I'_D is the expression in the table. This is only good for *very rough* calculations; see text.
[¶] See Fig. 2.14.
[¶¶] See Fig. 2.13.

Numerous footnotes are included in the tables. Trying to read these footnotes carefully at this time might be tiring and confusing. They can be passed over lightly for now, just so that the reader becomes aware of their existence. Their content will be appreciated later, when the reader has an overview of the basic aspects of modeling. The foot-notes can be found especially useful in the course of circuit design, when a particular problem happens to require special attention to the fine points that are discussed in them.

---

work easier with formulas that are used often and to get a "feel" for the magnitudes involved, one can put such formulas in a more convenient form by normalizing the quan-tities to commonly used units and even "reference" magnitudes. This is what we have done in the rightmost column of Table 3.3.

TABLE 3.2 V_{SB}-Dependent Quantities (nMOS Transistor)

$$V_T = V_{T0} + \gamma(\sqrt{V_{SB} + \phi_0} - \sqrt{\phi_0})\ *$$

$$V_M = V_T - cn\phi_t\ ^\dagger$$

$$I_M' = K'(n-1)\phi_t^2\ ^\ddagger$$

$$V_H \approx V_T + 6n\phi_t\ ^\S$$

$$I_H' \approx 20K'n\phi_t^2\ ^\S$$

$$n = 1 + \frac{\gamma}{2\sqrt{V_{SB} + \phi_0}}\ ^\P$$

$$\alpha = 1 + \frac{\gamma}{2\sqrt{V_{SB} + \phi_0 + \phi_\alpha}}\ ^{\P\P}$$

* For practical, ion-implanted devices, V_{T0} and ϕ_0 will be assumed given. Although it is common to calculate these quantities by using formulas developed for idealized, unimplanted devices (App. A), such practice should be used only as a last resort. A typical value for ϕ_0 is 0.7 V.

† c is a small number (e.g., 1 to 3). Its exact value depends on fabrication and V_{SB}, and it is usually not accurately known. Circuit design should never rely on the exact value of c. The quantities V_M and I_M can be determined from measurements by using the construction indicated in Fig. 2.12b. They are assumed to be chosen so that immediately below them, weak inversion expressions (including those for small-signal parameters in Chap. 4) yield acceptable error.

‡ This value is very approximate. Ideally I_M' should be determined from measurements for a known W/L value. See previous note.

§ Both V_H and $I_H = (W/L)I_H'$ fail to appear in the I_D expressions in Table 3.1. Their only role is to indicate lower limits for V_{GS} and $I_{D,SAT}$, below which strong inversion expressions, especially for small-signal quantities (Chap. 4), produce significant error. The values given for these limits are for approximate calculations. For precision work, the values of the limits may have to be increased.

¶ The formula for n assumes a modern fabrication process with negligible insulator-semiconductor *interface traps*. If this is not the case, the value of n can be somewhat higher. The formula also neglects a weak dependence on V_{GS} (Sec. 3.4).

¶¶ The quantity ϕ_α depends on bias, but is often assigned a single value that represents a compromise over the bias ranges of interest. For common applications with the terminal voltages between 0 and 5 V, a good compromise value for ϕ_α is 1 V. If the maximum value of the terminal voltages is restricted further, the optimum value of ϕ_α will be smaller, and it is sometimes chosen equal to 0; in that case, α reduces to n.

We will now "walk through" the three tables to explain the various quantities involved and to give a "feel" for the relations and their correct use.

3.2.1   The current equations

Let us first concentrate on the drain current equations in Table 3.1. The regions of inversion are defined there in terms of V_{GS}. An equivalent definition is also given in terms of $I_{D,SAT}$, which is the saturation current for a given V_{GS} value. Both types of definitions will be found useful in circuit work. The symbols V_M, I_M, V_H, and I_H used in these definitions were encountered in Figs. 2.12 to 2.14.

Weak inversion. Consider first the *weak* inversion equation in the table (a variation of the model presented in Ref. 5). The meaning of

TABLE 3.3   Model Parameters Expressed in Terms of Physical and Process Parameters*

Model parameter	Expression	Convenient formula[†]
ϕ_t	$\dfrac{\hat{k}T}{q}$	$25.9 \text{ mV} \times \dfrac{T}{300 \text{ K}}$
K'	$\mu C'_{ox}$	$34.5 \ \mu\text{A/V}^2 \times \dfrac{\mu}{100 \text{ cm}^2/(\text{V}\cdot\text{s})} \times \dfrac{100 \text{ Å}}{d_{ox}}$
C'_{ox}	$\dfrac{\epsilon_{ox}}{d_{ox}}$	$3.45 \text{ fF}/\mu\text{m}^2 \times \dfrac{100 \text{ Å}}{d_{ox}}$
γ	$\dfrac{\sqrt{2q\epsilon_s N_B}}{C'_{ox}}$	$0.168 \text{ V}^{1/2} \times \sqrt{\dfrac{N_B}{10^{16} \text{ cm}^{-3}}} \times \dfrac{d_{ox}}{100 \text{ Å}}$
V_A	—	$\phi_A \times \dfrac{L}{1 \ \mu\text{m}} \times \sqrt{\dfrac{N_C}{10^{16} \text{ cm}^{-3}}}$ [‡]

* Here $\hat{k}$ is Boltzmann's constant (1.38×10^{-23} V·C/K); T is the absolute temperature; the unit K means degrees Kelvin; q is the electron charge (1.602×10^{-19} C); μ is the carrier mobility in the channel; ϵ_{ox} is the permittivity of the insulator (3.45×10^{-13} F/cm for silicon dioxide); d_{ox} is the insulator thickness; ϵ_s is the semiconductor permittivity (1.04×10^{-12} F/cm for silicon); N_B is the bulk doping concentration; N_C is the doping concentration in the channel (which is usually different from N_B, since the channel is implanted).
† A silicon semiconductor and a silicon dioxide insulator are assumed in the relevant formulas.
‡ The value of ϕ_A depends on fabrication details; a compromise value is 16 V.

the quantities V_M and I_M can be understood with reference to Fig. 2.12. There V_M is the value of V_{GS} at the "top" of the weak inversion region, and I_M is the maximum weak inversion current, attained with $V_{GS} = V_M$ in saturation. As shown in the rightmost column of Table 3.1, I_M consists of two factors: the *aspect ratio* of the channel W/L (Fig. 2.1) and a process-dependent factor I'_M (typically 5 to 50 nA). The other quantities in the weak inversion current equation are the *thermal voltage* ϕ_t (25.9 mV at room temperature—see Table 3.3) and a quantity n, larger than 1 and usually less than 2. Plots obtained by using the weak inversion equation can look as shown in Fig. 3.1. As expected from the weak inversion equation in Table 3.1, saturation is approached when V_{DS} reaches values of a few times ϕ_t. This behavior is independent of V_{GS}, as seen from the factor in brackets in that table. For a plot using a logarithmic current axis, see the weak inversion part of Fig. 2.9a.

Example 3.1   It is given that $W = 9 \ \mu\text{m}$, $L = 3 \ \mu\text{m}$, $I'_M = 10 \text{ nA}$, and $n = 1.20$. Can a current of 460 pA be provided in weak inversion? If so, how far below the upper weak inversion limit should V_{GS} be, in order to obtain such a current in saturation?

solution   The maximum weak inversion current in saturation is $I_M = (W/L)I'_M = 30$ nA, so the given current value is safely in weak inversion. Using the current

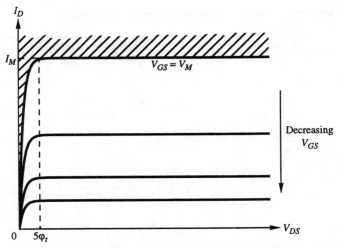

Figure 3.1 Weak inversion I_D-V_{DS} characteristics, with V_{GS} as a parameter and equal V_{GS} steps. The upper limit of weak inversion is at $V_{GS} = V_M$.

equation and neglecting the last exponential in it (since the device is in saturation), we obtain

$$V_{GS} - V_M = n\phi_t \ln \frac{I_D}{I_M} = 1.20(0.026 \text{ V}) \times \ln \frac{0.46 \text{ nA}}{30 \text{ nA}} = -0.13 \text{ V}$$

Thus V_{GS} should be 0.13 V below the upper limit of weak inversion in order to cause a current of 460 pA.

The above problem has been solved without our knowing the value of V_M. Note that one *cannot* use a rough value for V_M in the weak inversion current equation and hope to get a reasonable estimate for I_D, since V_M appears in an exponential and a slight error in its value can result in a large error in I_D.* This is illustrated in the next example.

Example 3.2   Problems caused by inaccuracies in the value of V_M.   Assume that, for the device of Example 3.1, it is also known that $V_M \approx 0.8$ V. Can the current at $V_{GS} = 0.7$ V and $V_{DS} = 2$ V be predicted with any reasonable accuracy?

solution   The answer is *no*. The value of V_M is only known approximately, to one decimal place; let us assume that the exact value could be anywhere from 0.75 to 0.85 V. Using the current equation, we obtain, for the given V_M value, $I_D = 1.2$ nA, whereas using instead 0.75 and 0.85 V, we obtain 6 and 0.24 nA, respectively. Thus an error of 400 percent is possible!

---

* The factor $I_M \exp[-V_M/(n\phi_t)]$ in the weak inversion equation is sometimes given as a single quantity, which we will denote by I_{DC}; the weak inversion current expression then becomes $I_{DC}\exp[V_{GS}/(n\phi_t)][1 - \exp(-V_{DS}/\phi_t)]$. This doesn't of course correct the problem; rather, the uncertainty in the value of V_M is now reflected in a large uncertainty in the value of I_{DC}.

The weak inversion equation can be useful besides the above problem, if one utilizes not V_M by itself but rather the quantity $V_{GS} - V_M$ as a whole, as was done in Example 3.1. Clever techniques enable the design of reliable circuits operating in weak inversion, despite uncertainties as to the exact value of V_M.* Such techniques rely on setting the bias *current,* rather than the bias voltage.

Moderate inversion.   The middle entry in Table 3.1 states the unfortunate fact that there is no simple expression available for I_D in *moderate* inversion. For quantitative results, in this region one has to use general models valid in all regions of inversion, which are, of course, more complicated (Sec. 3.5). Computer help is desirable. The reader should be warned, though, that some of the computer simulation models in wide use ignore the moderate inversion region altogether, and they assume that weak and strong inversion are adjacent regions. While the resulting error in the value of the current is sometimes bearable, such practice can lead to very wrong results in the values of small-signal parameters, as we will see in Chap. 4.

Strong inversion.   Let us now concentrate on the *strong* inversion entries in Table 3.1 [6]. The value of V_{GS} at the "bottom" of the strong inversion region is denoted by V_H; see Fig. 2.13. In strong inversion a two-segment model is used, with separate equations for nonsaturation and saturation, as seen in the table. Parameters K and k are proportionality constants in nonsaturation and saturation, respectively. Each of these quantities consists of two factors, as shown in the rightmost column: the channel aspect ratio W/L encountered above and a process-dependent factor (K' and k' for nonsaturation and saturation, respectively).[†] All these quantities are proportional to the effective mobility μ and the oxide capacitance per unit area C'_{ox}, as seen in Table 3.3; the reason for this was discussed in Sec. 2.9. Typical values for K' are 30 to 80 $\mu A/V^2$ for nMOS devices.

Another important model parameter in the strong inversion equations is V_T; this is the extrapolated threshold voltage already discussed in Secs. 2.7 and 2.8. Note in Fig. 2.13 that strong inversion does not start until $V_{GS} = V_H$, which is above V_T by a few tenths of a volt. Thus,

---

* The uncertainty in V_M, together with the fact that the effect of V_{SB} is hidden in several quantities in the weak inversion equation, makes the equation inadequate in cases where an accurate evaluation of the effect of V_{SB} on I_D is needed. For example, with $V_{GB} = V_{GS} + V_{SB}$ fixed, it turns out that I_D is simply proportional to $\exp(-V_{SB}/\phi_t)$ (Sec. 3.4); this is impossible to see from the equation in Table 3.1.

† The reader is warned that in some other treatments a factor of ½ is introduced in the definition of some of or all the quantities K, k, K', and k'; the current equations are then modified accordingly.

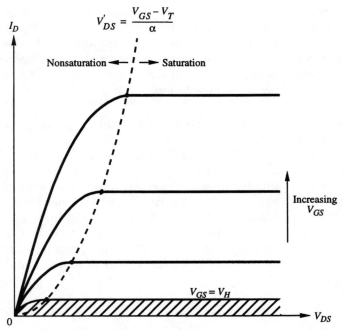

Figure 3.2 Strong inversion I_D-V_{DS} characteristics, with V_{GS} as a parameter. The lower limit of strong inversion is at $V_{GS} = V_H$.

when V_{GS} is equal to V_T (or only slightly above it), the device is *not* in strong inversion; nevertheless, the quantity V_T is a parameter encountered in the strong inversion relations, as seen in Table 3.1.* Finally, an additional parameter [6, 7, 1] used in the strong inversion relations is α, which usually lies between 1 and 1.5. The reader may have encountered the strong inversion model elsewhere with the value of α implicitly set to 1, in which case no distinction would be made between K' and k', or between K and k. This can lead to large errors, as will be seen shortly (Example 3.5).

Using the strong inversion equations, one obtains plots like those in Fig. 3.2. The equations for nonsaturation and for saturation produce the same I_D value at the limit between the two regions, which is denoted by V'_{DS}. The slope of the nonsaturation segment becomes zero at that point, and for larger values of V_{DS} the saturation equation takes over, predicting a current independent of V_{DS}. The nonsaturation equa-

---

* In very simple treatments, the transistor is assumed to be in strong inversion for any value of V_{GS} greater than V_T and "off" for V_{GS} less than V_T. Such simplifications lead to serious errors.

tion should never be used above $V_{DS} = V'_{DS}$, since it is then not valid and predicts meaningless behavior. As seen from both the equation for V'_{DS} in Table 3.1 and the plot in Fig. 3.2, the value of V_{DS} at which saturation is reached depends on the value of V_{GS}. This is in contrast to the case in weak inversion (see Fig. 3.1).

Example 3.3   Assume $K' = 77\ \mu A/V^2$, $V_T = 0.8$ V, $\alpha = 1.13$, $W = 9\ \mu m$, and $L = 3\ \mu m$. Find I_D with $V_{GS} = 3$ V and $V_{DS} = 0.5$ V. Assume that strong inversion begins at about $V_T + 0.2$ V.

solution   The strong inversion region starts at $V_H \approx V_T + 0.2$ V = 1 V. The given V_{GS} is above V_H, so we can use the strong inversion equations of Table 3.1. We have

$$V'_{DS} = \frac{3\ V - 0.8\ V}{1.13} = 1.95\ V$$

Since $V_{DS} < 1.95$ V, the device is in nonsaturation. We thus use the corresponding equation from Table 3.1, with $K = (W/L)K' = (9/3)(77) = 231\ \mu A/V^2$ to obtain

$$I_D = (231\ \mu A/V^2)\left\{\left[(3 - 0.8)0.5 - \frac{1}{2} \times 1.13 \times 0.5^2\right] V^2\right\} \approx 221\ \mu A$$

Example 3.4   In the above device, what value of V_{GS} is needed to produce a current of 60 μA in saturation?

solution   We have $k' = K'/\alpha = 68\ \mu A/V^2$, and $k = (W/L)k' = 204\ \mu A/V^2$. Assuming for the moment that the device operates in strong inversion, we obtain from the strong inversion saturation expression in Table 3.1

$$V_{GS} = V_T + \sqrt{\frac{2I_D}{k}} = 0.8\ V + \sqrt{\frac{2 \times 60\ \mu A}{204\ \mu A/V^2}} = 1.57\ V$$

It is now necessary to check our assumption of strong inversion operation. The bottom of strong inversion V_H is roughly at $V_T + 0.2$ V = 1 V. Thus the V_{GS} value found falls safely in strong inversion, and the result *is* valid.

Example 3.5   Error caused by assuming $\alpha = 1$.   In Fig. 3.3, curve 1 shows measured results for an nMOS transistor on a heavily doped substrate, taken from a commercial 14007 type of inverter [1]. If we use the strong inversion equations with $\alpha = 1$, we obtain curve 2, assuming a realistic value for K. The model succeeds deep in nonsaturation but clearly fails elsewhere, predicting a very large saturation current and an exaggerated V'_{DS} value. In an attempt to obtain better matching in saturation, K is sometimes artificially lowered. This results in curve 3. Now the model gets out of hand in nonsaturation, and it still predicts the same, exaggerated V'_{DS} value. It is thus obvious that using $\alpha = 1$, as is often done in circuit design texts, is inadequate even for approximate calculations for this device. In contrast, the model of Table 3.1 (with $\alpha = 1.7$) gives curve 4, which shows that a drastic improvement in accuracy results from including this parameter in the model. An intuitive feeling for the physical origin of this parameter will be provided shortly.

In Sec. 2.4.3 we saw that if the channel is not too long, the current in saturation can have a significant nonzero slope, due to the channel length modulation effect. This is illustrated by the solid curve in Fig. 3.4. The shape of this curve in saturation is a very complicated function

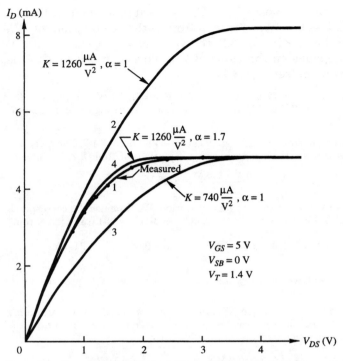

Figure 3.3 I_D versus V_{DS} for fixed V_{SB} and $V_{GS} = 5$ V. Curve 1: measured characteristic of an n-channel transistor on a heavily doped substrate (taken from one version of a commercial 14007-type CMOS inverter); curve 2: model with $\alpha = 1$, and other parameters chosen for good matching at low V_{DS} values; curve 3: model with $\alpha = 1$, with parameters chosen for good matching in the saturation region; curve 4: model with $\alpha = 1.7$, with parameters chosen for matching in both nonsaturation and saturation.

of V_{DS}, and its details also depend on V_{GS}, V_{SB}, the fabrication process, and the device geometry [1–4]. The equations discussed so far do not include this effect. For *very* rough estimates, the following modification is often used:

$$I_D \approx I_D'\left(1 + \frac{V_{DS}}{V_A}\right) \tag{3.3}$$

where I_D' is the value predicted for saturation by Table 3.1 and V_A is a quantity assumed to be independent of V_{DS}; this results in the straight-line approximation indicated by the broken line in Fig. 3.4. The quantity V_A is akin to the "Early voltage" used in bipolar device models, and it is chosen so that the slope has an appropriate value. The same factor $1 + V_{DS}/V_A$ is sometimes introduced also in the expression for the nonsaturation current. This is not strictly correct, since there is no channel

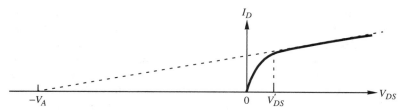

Figure 3.4 Graphical interpretation of the simple channel length modulation model.

length modulation in nonsaturation, but it is done to avoid a disconti-nuity which would otherwise occur at the boundary between nonsatu-ration and saturation; such a discontinuity could lead to problems (e.g., numerical difficulties in computer simulation). The resulting error in the nonsaturation current is often negligible. The value of V_A is smaller for devices with shorter channels and more lightly doped bodies since, as explained in Sec. 2.4.3, the saturation slope is larger for such devices. This is borne out in a formula for V_A given in Table 3.3.

Example 3.6 Reducing channel length modulation. Assume that when a partic-ular device operates in saturation with fixed V_{GS} and V_{SB}, its current varies by 2 percent over the V_{DS} range of interest. This is too large for a given application. How can the device be redesigned so that the variation is reduced to 1 percent? Assume that the formula for V_A in Table 3.3 is adequate.

solution Using Eq. (3.3), we find that if I_{D1} and I_{D2} are the current values at $V_{DS} = V_{DS1}$ and $V_{DS} = V_{DS2}$, respectively, we have $I_{D2} - I_{D1} = (V_{DS2} - V_{DS1})/V_A$. From the formula for V_A in Table 3.3, we see that V_A is proportional to L. Thus, to halve the variation, double the device length.

It should be emphasized again that the above approach is very rough, since it attempts to model too simply what is actually a very complicated phenomenon. After all, saturation curves, especially for some fabrication processes, are not even close to being straight lines. Thus the whole idea behind V_A is of limited value, unless one is willing to make V_A a function of bias.

The concept of the "pinch-off" region employed in Sec. 2.4.3 is, as men-tioned there, an oversimplification. Detailed analysis shows that the sit-uation near the drain region in saturation is actually very complicated [1–4]. The current flow is two-dimensional, with carriers dipping below the surface. Carrier velocity saturation and several short-channel effects (Sec. 3.8) have an additional effect on the shape of $I_D - V_{DS}$ curves in sat-uration. A phenomenon that can influence the drain current in the satu-ration region is *impact ionization*. In the "pinch-off" region, high-speed electrons can hit the crystal lattice atoms with enough energy to create electron-hole pairs. The electrons continue toward the drain, while the holes are collected by the substrate. This phenomenon causes a parasitic drain-substrate current I_{DB}, which increases strongly as V_{DS} is raised above V'_{DS}; it is observed as a nonzero substrate current, and it becomes a component of I_D, giving rise to an increasing slope of I_D with respect to V_{DS}. This effect is difficult to model.* With V_{DS} above V'_{DS} by 5 V, the above parasitic current can be on the order of 1 percent of I_D.

Detailed (and very complicated) models for the saturation region are discussed elsewhere [1–4, 6–19] and are found in many computer simulation programs. It should be emphasized, though, that for precision circuit work, even models used in some computer simulation programs are sometimes inadequate.

3.2.2 Model parameters influenced by the body effect

We have seen in Sec. 2.6 that an increase in the source-body bias V_{SB} causes a decrease in the drain current, assuming V_{GS} and V_{DS} remain unchanged. This is due to the body effect and is noticed in any region of inversion—compare Figs. 2.9 and 2.10. In the model equations of Table 3.1, the body effect is manifested in the fact that several model parameters are dependent on V_{SB}. These parameters are collected in Table 3.2.

The first entry in Table 3.2 gives the expression for the (extrapolated) threshold voltage V_T. In this expression, V_{T0} is the value of V_T when $V_{SB} = 0$ (Sec. 2.8), typically between 0.4 and 0.8 V for enhancement-mode devices in modern VLSI processes, and ϕ_0 is a characteristic potential with a typical value of 0.7 V. For realistic, ion-implanted devices, these quantities are difficult to calculate accurately [1]. For idealized, unimplanted devices, simple formulas do exist (App. A), but they should be used with practical devices only if no direct information on the values of the parameters in the threshold equation is available.

The parameter γ is the *body effect coefficient,* which determines how drastic the influence of V_{SB} is on V_T. This quantity is given in terms of physical and process parameters in Table 3.3. As we had anticipated in Sec. 2.4.2, the body effect is more pronounced for devices with heavily doped substrates and thick oxides. The value of γ usually lies between 0.2 and 1 $V^{1/2}$. A plot of V_T versus V_{SB} is shown in Fig. 3.5.†

The body effect also causes the V_{GS} limits between the regions of inversion to increase with V_{SB}, in a manner similar to that of V_T. Parameters V_M and V_H roughly track V_T, as indicated in Table 3.2. The uncertainty in the value of these quantities was discussed in Sec. 2.7. Again,

---

* An empirical relation [2] for the drain-substrate current is $I_{DB} \simeq \alpha_I I_D \exp[-V_I/(V_{DS} - 0.8V'_{DS})]$, where α_I and V_I are empirical constants dependent on the device details (typical values are $\alpha_I = 3$ and $V_I = 30$ V).

† For some implanted devices, different parameter values may have to be used in the V_T equation for different ranges of V_{SB} [1].

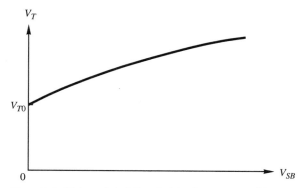

Figure 3.5 Extrapolated threshold voltage versus V_{SB}.

we caution the reader as to the proper use of the weak inversion current equation when V_M is not accurately known (Examples 3.1 and 3.2).

Finally, Table 3.2 gives also the dependence of I'_M, n, and α on V_{SB}. If γ is small, these quantities are sometimes taken to be fixed at their $0 - V_{SB}$ value (or some other compromise value) for simplicity.

From the formula for α we conclude that the common practice of setting $\alpha = 1$ (Example 3.5) cannot be justified unless the body effect coefficient is small and/or the source-body bias is large. The qualitative relation between α and γ can be explained intuitively. Recall that increasing the drain potential causes the channel to "pinch off," due to the manifestation of the body effect on the drain side of the channel (Sec. 2.4.3). If the body effect is large, relatively small values of the drain potential will be enough to cause pinch-off, and the current will not get a chance to "shoot" as high as it otherwise would—compare, in Fig. 3.3, the measured curve 1 to curve 2, which fails to take this effect into account. The tendency toward early saturation is manifested in the presence of α in the formula for V'_{DS} in Table 3.1; the tendency for lower saturation current is manifested in the presence of α in the formula for k' in the same table.

Considerations like the above are important not only for reasons of accuracy but also because they help reveal the correct *form* of the functional dependence of the current on the terminal voltages. In circuit design, correct knowledge of this form can be important.

We now give two examples of the use of Tables 3.2 and 3.3.

Example 3.7   Assume that for a given device $d_{ox} = 200$ Å, $\mu = 445$ cm^2/(V·s),* $N_B = 1.1 \times 10^{16}$ cm^{-3}, and $N_C = 1.7 \times 10^{16}$ cm^{-3}. Find C'_{ox}, K', γ, and V_A/L.

---

* This value may seem somewhat low in comparison to independent mobility measurements. Here we have an example of choosing an artificial value for a model parameter in order to improve the overall accuracy of a simple model. See the related comments in Sec. 3.2.3.

solution   From Table 3.3 we have

$$C'_{ox} = 3.45 \text{ fF/}\mu\text{m}^2 \times \frac{100}{200} \approx 1.73 \text{ fF/}\mu\text{m}^2$$

$$K' = 34.5 \text{ }\mu\text{A/V}^2 \times \frac{445}{100} \times \frac{100}{200} \approx 77 \text{ }\mu\text{A/V}^2$$

$$\gamma = 0.168 \text{ V}^{1/2} \times \sqrt{\frac{1.1 \times 10^{16}}{10^{16}} \times \frac{200}{100}} \approx 0.35 \text{ V}^{1/2}$$

$$\frac{V_A}{L} = 16 \text{ V} \times \frac{1}{1 \text{ }\mu\text{m}} \times \sqrt{\frac{1.7 \times 10^{16}}{10^{16}}} \approx 21 \text{ V/}\mu\text{m}$$

Example 3.8   A device has electrical parameter values $V_{T0} = 0.8$ V, $\phi_0 = 0.76$ V, $K' = 77 \text{ }\mu\text{A/V}^2$, $\gamma = 0.35 \text{ V}^{1/2}$, and $V_A/L = 21 \text{ V/}\mu\text{m}$ (the last three values were determined from process parameters in Example 3.7).

(a) Find V_T, α, k', n, and I'_M at $V_{SB} = 0$ V.
(b) Repeat part (a) for $V_{SB} = 2$ V.
(c) If $W = 10 \text{ }\mu\text{m}$ and $L = 3 \text{ }\mu\text{m}$, find K, k, and I_M at $V_{SB} = 2$ V. Also find V_A.

solution   (a) Using the given parameter values and $V_{SB} = 2$ V in the expressions of Table 3.2, we have

$$V_T = V_{T0} = 0.8 \text{ V}$$

$$\alpha = 1 + \frac{0.35 \text{ V}^{1/2}}{2\sqrt{(0 + 0.76 + 1.0) \text{ V}}} \approx 1.13$$

$$k' = \frac{77 \text{ }\mu\text{A/V}^2}{1.13} \approx 68 \text{ }\mu\text{A/V}^2$$

$$n = 1 + \frac{0.35 \text{ V}^{1/2}}{2\sqrt{(0 + 0.76) \text{ V}}} = 1.20$$

$$I'_M = (77 \text{ }\mu\text{A/V}^2)(1.20 - 1)(0.0259 \text{ V})^2 \approx 10 \text{ nA}$$

(b) Repeating with $V_{SB} = 2$ V, we obtain

$$V_T = 0.8 \text{ V} + 0.35 \text{ V}^{1/2} \times (\sqrt{(2 + 0.76) \text{ V}} - \sqrt{0.76 \text{ V}}) \approx 1.08 \text{ V}$$

$$\alpha = 1 + \frac{0.35 \text{ V}^{1/2}}{2\sqrt{(2 + 0.76 + 1.0) \text{ V}}} = 1.09$$

$$k' = \frac{77 \text{ }\mu\text{A/V}^2}{1.09} \approx 71 \text{ }\mu\text{A/V}^2$$

$$n = 1 + \frac{0.35 \text{ V}^{1/2}}{2\sqrt{(2 + 0.76) \text{ V}}} = 1.105$$

$$I'_M = (77 \text{ }\mu\text{A/V}^2)(1.105 - 1)(0.0259 \text{ V})^2 \approx 5.4 \text{ nA}$$

(c) Multiplying the given value of K' and the values of k' and I'_M from part (b) by W/L, we have

$$K = 77 \ \mu A/V^2 \times \frac{10}{3} \approx 257 \ \mu A/V^2$$

$$k = 71 \ \mu A/V^2 \times \frac{10}{3} \approx 237 \ \mu A/V^2$$

$$I_M = 5.4 \ \mu A \times \frac{10}{3} = 18 \ nA$$

Finally, multiplying the given value of V_A/L by the given L, we have

$$V_A = 21 \ V/\mu m \times 3 \ \mu m = 63 \ V$$

3.2.3 Origin and validity of the model

The model just presented is a tried-and-tested combination of several results [5, 6, 1], developed originally for idealized, uniform-substrate MOS transistors. Strictly speaking, then, the model should be expected to be valid only for such "academic" devices, especially if "ideal" values are used for the various parameters in it. For many applications, though, the model can be "stretched" to cover real, implanted devices if one allows some of the parameters to be adjusted for best fit to measurements; then the model can often prove reasonable for approximate calculations, at least over limited bias ranges. Thus, e.g., the strong inversion I_D equations of Table 3.1 do not have the exact functional form needed for implanted devices, but they may be able to approximately model such devices over bias ranges of a couple of volts, if the parameters K', V_T, and α can be "extracted" from measurements done in those ranges (Sec. 3.12). This is normally not practical, however, since it necessitates a different extraction for each V_{SB} value of interest. One then uses the expressions given in Tables 3.2 and 3.3 for these three parameters, but at least the values of μ, V_{T0}, γ, and ϕ_0 found in these expressions *should* be extracted from measurements. The values of these parameters after such extraction may be "nonphysical"; e.g., the value of the mobility may turn out to be lower than what would be expected from independent mobility measurements.

The accuracy in modeling implanted devices can be significantly affected by the fact that the expression for V_T does not have the exact functional form needed for such devices; so it cannot predict V_T accurately over a large range of V_{SB} regardless of what parameter values are used in it. At least one can minimize the error over a given V_{SB} range, if the parameters are well chosen. If even the above extraction is not possible, one may have to use "ideal" or typical values for the parameters; but in that case there should be no illusions as to the accuracy that can be expected.

Note that models developed explicitly for implanted devices do exist [1–4]. They are too complicated for hand analysis, though. One would at least hope to find such models in computer simulators, but unfortunately this is often not the case. Of course, good circuit design almost never relies on accurate values of voltages and currents, since such values cannot be guaranteed owing to fabrication tolerances, temperature variations and other environmental factors, and aging. Sometimes, though, the uncertainty due to poor modeling exceeds the uncertainty due to such factors (Sec. 3.12). A circuit designer is thus forced to design exceedingly conservatively, to ensure that the circuit will work when it is manufactured.

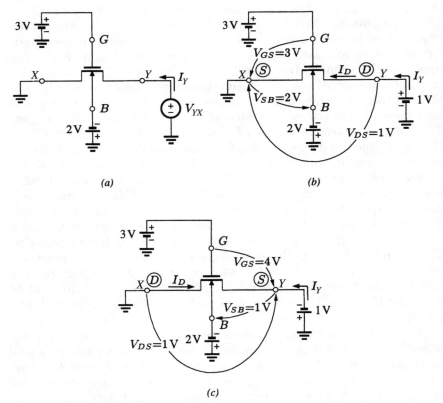

Figure 3.6 Identification of source and drain terminals is based on the polarity of the total channel voltage.

3.3 Drain versus Source

In writing the model equations above, we have assumed, as stated in Table 3.1, that for the two terminals labeled D and S the following holds:

$$V_{DS} \geq 0 \qquad (3.4)$$

If in a given situation this does not hold, the labels of the terminals must be interchanged before the equations can be applied.

Example 3.9 For the device in Fig. 3.6a (where a common symbol for nMOS transistors is used), assume $K = 60 \ \mu A/V^2$, $V_{T0} = 1$ V, $\phi_0 = 0.8$ V, and $\gamma = 0.5 \ V^{1/2}$. Calculate the current I_Y for (a) $V_{YX} = 1$ V and (b) $V_{YX} = -1$ V.

solution (a) With $V_{YX} = 1$ V > 0, the drain is on the right and the source is on the left, as shown in Fig. 3.6b.* Thus the various terminal potential differences are as indicated in the figure. From Tables 3.1 and 3.2 we then find $V_T = 1.39$ V, $\alpha = 1.13$, and $I_D \simeq 63 \ \mu A$. Since $I_Y = I_D$, we have

$$I_Y \simeq 63 \ \mu A$$

(b) With $V_{YX} = -1$ V < 0, the roles of the X and Y terminals are reversed; now X is the drain, and Y is the source, as shown in Fig. 3.6c. Using the voltages shown in this figure, we find $V_T = 1.22$ V, $\alpha = 1.15$, and $I_D \simeq 132 \ \mu A$. Now $I_Y = -I_D$, and

$$I_Y \simeq -132 \ \mu A$$

---

* In this book, a potential of a point U with respect to a point Z will be indicated by an arrow pointing from U to Z.

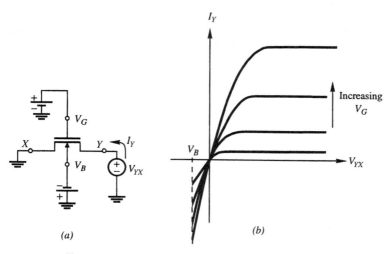

Figure 3.7 Channel current versus total channel voltage, with the latter allowed to assume values of either polarity.

In addition to the sign reversal, note that the magnitude of the second current is different, for *two* reasons: The gate-source voltage is larger, and the threshold voltage is smaller due to the smaller source-substrate voltage.

The structure we used in the above example is repeated in Fig. 3.7, along with a set of characteristics. The parameter V_G is defined with respect to *ground,* and *not* with respect to the "source." The characteristics can be calculated as illustrated in Example 3.9. Note that V_{YX} cannot be reduced below V_B, for then the internal *pn* junction corresponding to terminals *B* and *Y* would become forward-biased.

3.4   Symmetric Models

MOS transistors are not always laid out symmetrically. For example, the source area may surround the gate and drain areas, in order to keep the chip area small for a transistor with a large *W,* or to minimize parasitic drain capacitances. Examples of such layout will be seen in Chap. 6. High-order effects present in nonsymmetric structures can make the source and drain not perfectly interchangeable. If the two terminals are interchanged, slightly different parameter values may be needed to describe the transistor characteristics well.

Nevertheless, MOS transistors often *are* laid out and made symmetrically, as illustrated in simplified form in Fig. 2.1. In such cases, it is sometimes useful to have models which clearly exhibit the interchangeability between source and drain, and the corresponding equivalence between the source and drain potentials with respect to a common reference, conveniently taken as the body, as shown in Fig. 3.8. We will now present such models.

Strong inversion.   An accurate symmetric model for the *n*MOS transistor in the strong inversion *nonsaturation* region is [20, 21, 1]:

$$I_D = K \left\{ (V_{GB} - V_{T0} + \gamma\sqrt{\phi_0})(V_{DB} - V_{SB}) - \frac{1}{2}(V_{DB}^2 - V_{SB}^2) \right.$$

$$\left. - \frac{2}{3}\gamma[(V_{DB} + \phi_0)^{3/2} - (V_{SB} + \phi_0)^{3/2}] \right\}, \qquad \text{nonsaturation} \qquad (3.5)$$

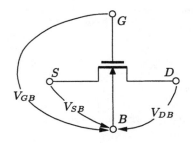

Figure 3.8 Terminal voltages defined with respect to the body.

where V_{GB} is the gate-body voltage, V_{SB} the source-body voltage, and V_{DB} the drain-body voltage. Saturation is reached when V_{DB} reaches a value V_P which pinches off the drain end of the channel, and which is given by (Prob. 3.18)

$$V_P = V_{GB} - V_{T0} - \gamma \left[\sqrt{V_{GB} - V_{T0} + \left(\frac{\gamma}{2} + \sqrt{\phi_0} \right)^2} - \left(\frac{\gamma}{2} + \sqrt{\phi_0} \right) \right] \quad (3.6)$$

Equation (3.5) is not valid in the saturation region; the model can be extended into saturation by using an approach similar to the one used earlier (Prob. 3.18).

Since the device is symmetric, the roles of drain and source can be interchanged. Thus, if V_{DB} does not reach V_P but instead V_{SB} does, we will have pinch-off at the *source* end of the channel and the current will saturate again (flowing in the opposite direction from before). This is termed *reverse* saturation. Due to its complete symmetry, the model could have been used to calculate the two currents in Fig. 3.6b and c without having to interchange the labels of the two terminals. It is interesting to point out that our strong inversion model in Table 3.1 can be derived from the above symmetric model (Prob. 3.19).

A simplified version of the above symmetric model, adequate for approximate results in *low-voltage* work (in which the possible ranges of the terminal voltages are limited), can be obtained by appropriate series expansion of Eq. (3.5) and is [22] (Prob. 3.20)

$$I_D = K \left[(V_{GB} - V_{T0})(V_{DB} - V_{SB}) - \frac{1}{2} m_0 (V_{DB}^2 - V_{SB}^2) \right], \quad \text{nonsaturation}$$

$$(3.7)$$

where nominally

$$m_0 = 1 + \frac{\gamma}{2 \sqrt{\phi_0}} \quad (3.8)$$

but this constant can be adjusted for best overall fit to measurements, once the possible range of the terminal voltages is known. For the model of Eq. (3.7), saturation (or reverse saturation) is obtained when V_{DB} (or V_{SB}) reaches the following "pinch-off" value (Prob. 3.20):

$$V_P = \frac{V_{GB} - V_{T0}}{m_0} \quad (3.9)$$

In saturation, Eq. (3.7) is no longer valid. The saturation region current can be obtained by extension from the current at the upper end of nonsaturation, as before. For example, using $V_{DB} = V_P$ from Eq. (3.9) in Eq. (3.7), we obtain (Prob. 3.20)

$$I_D = \frac{K}{2m_0} (V_{GB} - V_{T0} - m_0 V_{SB})^2, \quad \text{saturation} \quad (3.10)$$

which is related to the strong inversion saturation expression of Table 3.1 (Prob. 3.20).

Weak inversion.   For the weak inversion region, a symmetric model is [23–28]

$$I_D = 2Km\,\phi_t^2 \exp\!\left(\frac{V_{GB} - V_{T0}}{m\,\phi_t}\right)\!\left[\exp\!\left(-\frac{V_{SB}}{\phi_t}\right) - \exp\!\left(-\frac{V_{DB}}{\phi_t}\right)\right] \quad (3.11)$$

where ϕ_t is the thermal voltage (Table 3.3) and m is larger than 1 and usually less than 2. The value of m is sometimes considered constant and, for simplicity, equal to m_0 from Eq. (3.8); for more accurate work, the following value can be used [23–28]:

$$m = \left[1 - \frac{\gamma}{2\sqrt{V_{GB} - V_{T0} + (\gamma/2 + \sqrt{\phi_0})^2}}\right]^{-1} \quad (3.12)$$

Equations (3.11) and (3.12) are a more accurate weak inversion model than the one in Table 3.1. This accurate model reveals the fact that I_D is not an exact exponential with respect to the gate voltage, since m also depends on the latter (nevertheless, this dependence is weak and is hardly noticeable in log I_D versus V_{GS} plots, such as that in Fig. 2.12b). Also, the above model depicts the dependence of I_D on V_{SB} accurately, in contrast to the expression in Table 3.1, in which V_{SB} was hidden in various quantities dependent on it. The dependence on V_{SB} is now seen from Eq. (3.11) to be purely exponential, as long as V_{GB} is held constant.

Despite the appealing symmetries of the above models, and the fact that in certain cases they offer advantages, most circuit designers prefer to work with the source-referenced models we have presented earlier (partly due to convenience and partly due to habit—emitter-referenced models are standard in bipolar transistor circuit design). This will be our preference in this book, too, except in certain special cases.

3.5   General Models and Moderate Inversion

A comparison of our simple I_D expressions of Sec. 3.2 to accurate I_D values is shown in Fig. 3.9. As seen, both the weak and strong inversion expressions become very inaccurate in moderate inversion [29]. In fact, it can be calculated [23–28] that at point Z, these expressions predict a current several times larger and smaller, respectively, than the actual device current! (The I_D axis is logarithmic in the figure.)

Despite the obvious failure of strong and weak inversion expressions in the moderate inversion region, some models in common use (even in popular computer simulators, for example, Spice level 2 and level 3 models) ignore the moderate inversion region and assume that the

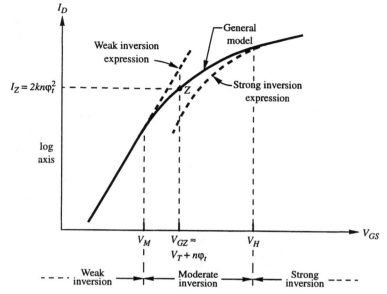

Figure 3.9   Drain current from measurements (solid line) compared to weak and strong inversion expressions (broken lines).

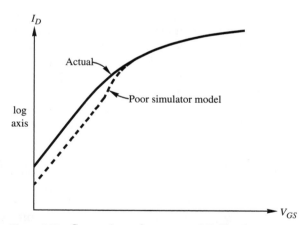

Figure 3.10   Comparison of a measured I_D-V_{GS} characteristic to that obtained from a common simulator model.

weak and strong inversion regions are adjacent at a point, often taken as $V_{GS} = V_T + n\phi_t$. To enable current continuity, the weak and strong inversion curves are artificially shifted to make them touch at this point, as shown in Fig. 3.10. This not only sacrifices accuracy, but also results in an unnatural kink, as seen. The use of artificial values for the model parameters can reduce the magnitude of the problem to some extent, but it can lead to other complications (Sec. 3.12). The kink mentioned creates very serious inaccuracies in the values of small-signal parameters, as we shall see in Chap. 4.

It is thus obvious that the moderate inversion current must be modeled accurately, especially since many devices in today's low-voltage, low-power circuits operate in this region [23, 25, 27]. Unfortunately, no simple model is available for the moderate inversion region, so one must resort to general models, which are valid for any combination of bias voltage values. It is, in fact, possible to model the drain current by using a single expression which is continuous and valid in all regions of inversion (weak, moderate, and strong). When derived from physical considerations, though, such models are complicated and do not provide the current as an explicit function of the terminal voltages [30, 1]. Things become simpler if empirical interpolation functions are used. An example of a model using such functions is [24–28]:

$$I_D = 2Km\phi_t^2\left\{\ln^2\left[1 + \exp\left(\frac{V_{GB} - V_{T0} - mV_{SB}}{2m\phi_t}\right)\right]\right.$$

$$\left. - \ln^2\left[1 + \exp\left(\frac{V_{GB} - V_{T0} - mV_{DB}}{2m\phi_t}\right)\right]\right\} \quad (3.13)$$

with m given by Eq. (3.12) and ln denoting the natural logarithm. This "single-piece" model can be used in all regions, for both nonsaturation and saturation. It gives a smooth curve like the solid line in Fig. 3.9, which approaches asymptotically the weak and strong inversion models in the respective regions. In weak inversion both exponentials have a magnitude much smaller than 1; then by using the approximation $\ln(1 + x) \simeq x$ for $|x| \ll 1$, it is easily seen that the model reduces to the weak inversion symmetric model of Eq. (3.11). In strong inversion and deeply in nonsaturation, both exponentials are much larger than 1. Using now the approximation $\ln^2(1 + e^y) \simeq \ln^2 e^y = y^2$ for $e^y \gg 1$, we see that these exponentials produce square-law terms and the model reduces to Eq. (3.7) with m_0 replaced by m. In moderate inversion no simplification is possible, and the entire expression (3.13) must be used.* As V_{DB} is raised, the second exponential in Eq. (3.13) becomes negligible; saturation is then naturally obtained, and in strong inversion the equation reduces to Eq. (3.10) with m_0 replaced by m (Prob. 3.21). If V_{SB} is raised instead, then it is the first exponential that

---

* In fact, in computer simulators the entire expression should be used throughout, to avoid abrupt transitions in the boundaries between the regions, which could cause numerical difficulties and would result in severe error in the small-signal parameters (Chap. 4).

becomes negligible and we enter reverse saturation. Several other single-piece models are described in the literature [11–14, 31].

One can extend the above ideas to a model that uses voltages referenced to the source. Using the same type of interpolation as Eq. (3.13) [24–28], but modifying the expression so that it reduces to familiar results [6] in strong inversion, we obtain the model summarized in Table 3.4 [32]. This model makes explicit use of the threshold voltage V_T, given for each V_{SB} value by the familiar expression shown in the table (this expression is from Table 3.2 and is repeated here for convenience). Since V_T is explicitly used in the model, if desired, one can use for it values found from more elaborate V_T models, e.g., to take into account effects related to ion implantation [1] or to small channel dimensions (Sec. 3.8). The model in Table 3.4 is valid in all regions, including moderate inversion. In weak inversion, with both exponentials much smaller than 1 and, as before, $\ln(1 + x) \simeq x$ for $|x| \ll 1$, the model can be shown to reduce to (Prob. 3.22)

$$I_D \simeq I_Z \exp\left(\frac{V_{GS} - V_T}{n\phi_t}\right)\left[1 - \exp\left(-\frac{V_{DS}}{\phi_t}\right)\right], \qquad \text{weak inversion} \qquad (3.14)$$

which is consistent with the weak inversion result in Table 3.1.*

---

* For the latter to be asymptotically consistent with Eq. (3.14), one must use V_M as given in Table 3.2, with $c = \ln\left[2n/(n-1)\right]$ (Prob. 3.22).

TABLE 3.4 Large-Signal DC Model for Drain Current in All Regions (nMOS Transistor)

$$I_D = I_Z \left\{ \ln^2\left[1 + \exp\left(\frac{V_{GS} - V_T}{2n\phi_t}\right)\right] - \ln^2\left[1 + \exp\left(\frac{V_{GS} - V_T - nV_{DS}}{2n\phi_t}\right)\right] \right\}$$

where

$$I_Z = 2Kn\phi_t^2$$

$$K = \frac{W}{L}K'$$

$$V_T = V_{T0} + \gamma(\sqrt{V_{SB} + \phi_0} - \sqrt{\phi_0})$$

$$n = 1 + \frac{\gamma}{2\sqrt{V_{SB} + \phi_0}}$$

NOTES: The model [32] has been adopted from the model in Refs. 24 to 28 for using the source as a reference and for making possible the explicit inclusion of an accurate V_T value at any V_{SB} value.

$V_{DS} \geq 0$ is assumed. The model is valid in weak, moderate, and strong inversion, nonsaturation and saturation. Junction leakage and breakdown effects are not included.

If it is desired to include the dependence of I_D on V_{DS} in saturation, multiply by $1 + V_{DS}/V_A$ (see Sec. 3.2).

Quantities K', γ, ϕ_t, and V_A are given in Table 3.3. The values of V_{T0} and ϕ_0 are assumed known for a given process. A typical value of ϕ_0 is 0.7 V. An improvement is possible if ϕ_0 is replaced by a varying potential; see footnote in text on page 70.

Consider now strong inversion. Here the first exponential in Table 3.4 is much larger than 1. If V_{DS} is very small (i.e., we are in deep nonsaturation), the second exponential is much larger than 1, too. Using then the approximation $\ln^2(1 + e^y) \simeq \ln^2 e^y = y^2$, valid for $e^y \gg 1$, and after some algebra, we obtain (Prob. 3.22):

$$I_D \simeq K\left[(V_{GS} - V_T)V_{DS} - \frac{1}{2}\, nV_{DS}^2\right], \qquad \begin{array}{c}\text{strong inversion---}\\ \text{deep nonsaturation}\end{array} \qquad (3.15)$$

If, in strong inversion, V_{DS} is raised, the second exponential in the current expression of Table 3.4 becomes negligible, whereas the first exponential remains much larger than 1. We are then in saturation, and the expression simplifies to (Prob. 3.22)

$$I_D \simeq \frac{K}{2n}\,(V_{GS} - V_T)^2, \qquad \text{strong inversion---saturation} \qquad (3.16)$$

It can be seen that Eqs. (3.15) and (3.16) are the same as the corresponding strong inversion expressions in Table 3.1, with α replaced by n.*

In saturation (for any region of inversion) the second exponential is negligible, and we have

$$I_D = I_Z \ln^2\left[1 + \exp\left(\frac{V_{GS} - V_T}{2n\,\phi_t}\right)\right], \qquad \text{saturation, all regions of inversion}$$
$$(3.17)$$

The above saturation current is approached asymptotically as V_{DS} is increased; there are no artificial seams between nonsaturation and saturation.

A comparison of the models of Eq. (3.13) and of Table 3.4 to measured results is shown in Fig. 3.11 (both models produce practically identical results for $V_{SB} = 0$) [32]. It is seen that not only are there no kinks of the type illustrated in Fig. 3.10, but the accuracy is good throughout.

3.6   Mobility Dependence on Gate and Substrate Bias

The mobility of electrons in the inversion layer is lower than that in bulk silicon, because the gate field pulls the inversion layer tightly toward the

---

* The replacement of α by n is acceptable in low-voltage work. In cases where the terminal voltages must vary over a significant range (for example, 0 to 5 V) and a small error is desired, the accuracy of the model in Table 3.4 can be improved by allowing n to change gradually into α, as V_{GS} is raised and strong inversion is entered. One way to do this is to replace ϕ_0 in the formula for n by a quantity which is a function of V_{GS} and to make this quantity change from ϕ_0 in weak inversion to $\phi_0 + \phi_\alpha$ (Table 3.2) in strong inversion [32].

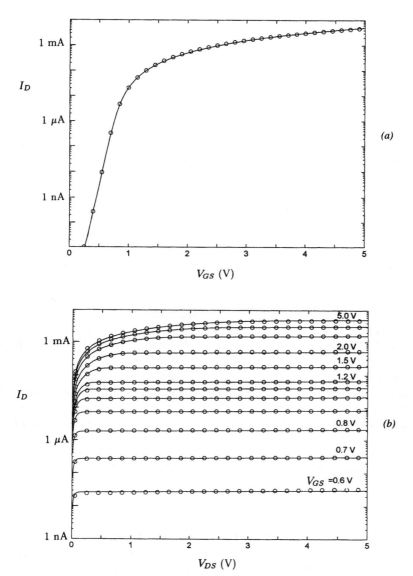

Figure 3.11 Comparison of the models of Eq. (3.13) [24–28] and of Table 3.4 to measurements for a long-channel device ($W = 108$ µm, $L = 7$ µm) [32]. (a) I_D (on a logarithmic axis) versus V_{GS}, with $V_{DS} = 5$ V and $V_{SB} = 0$ V; (b) I_D (on a logarithmic axis) versus V_{DS}, for $V_{SB} = 0$ V. (For $V_{SB} = 0$ V both models produce practically identical results.)

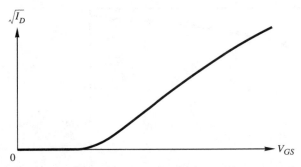

Figure 3.12 $\sqrt{I_D}$ versus V_{GS} in saturation. The top curved part is affected by effective mobility reduction; the bottom curved part is affected by moderate and weak inversion.

oxide-silicon interface and the electrons find it more difficult to move in it. This effect becomes more pronounced as V_{GS} is increased. For some devices, the effect can be easily observed on plots of $\sqrt{I_D}$ versus V_{GS} in the strong inversion saturation region. Such plots are supposed to observe the equation $\sqrt{I_D} = \sqrt{k/2}(V_{GS} - V_T)$ with $k = (W/L)\mu C'_{ox}/\alpha$ (see Tables 3.1 and 3.3), but instead of being straight lines as in Fig. 2.14b, they look as shown in Fig. 3.12; as seen, in strong inversion the slope decreases with increasing V_{GS} (the bottom part of the curve is not relevant here, as it corresponds to moderate and weak inversion). This effect can be modeled by making μ (and thus k' and k) a decreasing function of V_{GS}. As Fig. 3.12 implies, a device with V_{GS}-dependent mobility is not a true square-law device; the spacing between successive curves in saturation (Fig. 2.9b) will increase somewhat less drastically than square law would have it, with increasing V_{GS}. The value of V_{SB} can be shown to also affect the mobility in a similar manner (μ decreasing with increasing V_{SB}), although less strongly. Both effects can be modeled by [2]

$$\mu = \frac{\mu_0}{1 + \theta(V_{GS} - V_T) + \theta_B V_{SB}} \tag{3.18}$$

where μ_0 is the low-field mobility [typically 600 cm²/(V·s)], θ is a process-dependent quantity (with a typical value of 25 Å·V^{-1}/d_{ox}, where d_{ox} is the oxide thickness), and θ_B is typically a few hundredths of 1 V^{-1}. The value of θ is sometimes artificially increased to approximately model the effect of the source series resistance (see Sec. 3.8). The effect of V_{SB} is more important in p-channel devices. Note that omitting the last term in the denominator of the above equation, as is often done, will result in predicting that μ increases with increasing V_{SB} (through the increasing V_T), which is false. This problem is avoided by including the last term and choosing θ_B sufficiently large. Other formulas are also used [1].

3.7 Temperature Effects

Most model parameters we have encountered depend on temperature. Two have been extensively characterized in the literature: the mobility and the threshold voltage. The effective mobility is known to decrease with temperature; it can decrease from its room-temperature value by 15 to 20 percent at a temperature of 60°C. It can be modeled by

$$\mu(T) = \mu(T_r)\left(\frac{T}{T_r}\right)^{-c_1} \tag{3.19}$$

where T is the absolute temperature, T_r is room absolute temperature, and c_1 is a constant between 1.5 and 2.

The threshold voltage decreases practically linearly with temperature and can be approximated by

$$V_T(T) = V_T(T_r) - c_2(T - T_r) \tag{3.20}$$

where c_2 is usually between 0.5 and 4 mV/K (typically 2 mV/K).

Example 3.10 Assume that for satisfactory operation a circuit needs devices with a threshold voltage of at least 0.4 V. Let the threshold voltage be 0.5 V at 27°C, and assume it decreases by 2 mV/°C. Then a 50°C increase in temperature will lower the threshold voltage by 0.1 V; thus the highest temperature of operation for this circuit will be 77°C, assuming the temperature increase does not adversely affect operation in any other way.

Figure 3.13 shows a plot of $\sqrt{I_D}$ versus V_{GS} for different temperatures in saturation, where one expects (Tables 3.1 and 3.3) that, for strong inversion, $\sqrt{I_D} = \sqrt{k/2}(V_{GS} - V_T)$, with $k = (W/L)\mu C'_{ox}/\alpha$. A temperature increase tends to increase the drain current through $V_{GS} - V_T$ and to decrease it through μ. At high currents, the decrease through μ wins

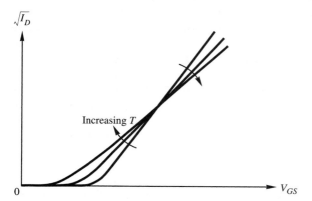

Figure 3.13 $\sqrt{I_D}$ versus V_{GS} in saturation for various temperatures. The bottom curved parts of the curves are due to moderate and weak inversion.

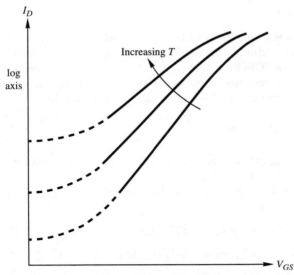

Figure 3.14 I_D (on a logarithmic axis) versus V_{GS} for low V_{GS} values in saturation, for various temperatures. The top curved parts of the curves are due to moderate and strong inversion. The bottom broken-line parts are influenced by leakage currents.

out; the opposite is true at low currents. In certain cases a value of V_{GS} can be found at which the current becomes practically temperature-independent over a large temperature range. This effect is evident in the figure. The bottom, curved part of the characteristic is due to moderate and weak inversion.

As can be deduced from Fig. 3.13, for low V_{GS} values and for a given V_{GS}, the drain current increases with temperature. Plots of the saturation I_D (on a logarithmic scale) versus V_{GS}, for low V_{GS} values, are shown for various temperatures in Fig. 3.14. As seen, with increasing temperature the slope of the curves decreases. Also, the junction leakage drastically increases with temperature; it can double for an 8 to 10°C temperature increase (the parts of the curves influenced by leakage are shown by broken lines in the figure). This leakage masks the weak inversion current, thus diminishing the range of currents over which exponential behavior is observed. For this reason it is difficult to design circuits operating in weak inversion at high temperatures.

3.8 Small-Dimension Effects

Small transistor dimensions are desirable because of the resulting advantages in circuit complexity per unit of chip area, fabrication cost per circuit of given complexity, speed of operation, and circuit power dissipation. When a fabrication process is used to make devices of very small dimensions (near the limits of its capability), the models we have presented in Tables 3.1 to 3.4 can be of limited value unless they are modified. We will use the term *small-dimension effects* for various phenomena responsible for this problem, and we will briefly outline these phenomena below; we will also discuss what modifications should be performed on our models to stretch their validity to such cases. More quantitave results can be found elsewhere [1–4].

1. When the channel length L becomes small, the electric field produced by the drain and source can influence directly much of the channel and the depletion region underneath it. In particular, part of the role of depleting the region under the channel is now undertaken by the source and drain; the gate field instead, being relieved of this task in part, goes to invert the channel more strongly. This fact manifests itself as a threshold voltage reduction; thus V_{T0} *decreases when L is decreased.* A few hundreds of millivolts of reduction are possible when L is reduced to the minimum value allowed by reliability considerations (a fraction of a micrometer in modern processes).

2. The above effect becomes more pronounced as the drain depletion region becomes wider, which will happen as V_{DS} is increased. Thus, for a fixed small L, V_{T0} *decreases as V_{DS} is increased.* The decrease in V_{T0} is roughly proportional to V_{DS} and continues even above $V_{DS} = V'_{DS}$, thus increasing $V_{GS} - V_T$ and contributing to the nonzero slope of I_D versus V_{DS} in saturation. The decrease in V_{T0} can be a few tenths of a volt for a V_{DS} of several volts, when L is near its minimum allowed value. The phenomenon just described is often explained by the concept of *drain-induced barrier lowering,* according to which an increase in V_{DS} lowers the barrier to the electrons entering the channel from the source.

3. Because in short-channel (small-L) devices part of the channel charge is controlled by the fields of the drain and source, the influence of the body bias becomes smaller. Thus, *the effective body effect coefficient decreases when L is decreased.*

4. For a fixed L, if the gate width W is reduced to near its minimum value, a nonnegligible part of the field lines emanating from the gate "fringe" outside the area immediately below it; the field under the gate is thus left weaker, and the inversion is less heavy, which manifests itself as an increase in the threshold voltage. So V_{T0} *increases when W is decreased.*

5. In narrow-channel (small-W) devices, the gate field depletes a wider body region due to the aforementioned fringing field. Thus more body charge figures in the charge balance, and the influence of the body becomes larger. *The effective body effect coefficient becomes larger when W is decreased.*

The above five effects are difficult to model accurately. For first-order results *over limited ranges* of L, W, and V_{DS}, one can use [1]

$$V_T = V_T' - a_l \frac{d_{ox}}{L}(V_{SB} + \phi_0) - a_v \frac{d_{ox}}{L} V_{DS} + a_w \frac{d_{ox}}{W}(V_{SB} + \phi_0) \qquad (3.21)$$

where V_T' is the threshold voltage of a large-dimension device as calculated from Table 3.2 (including the body effect), d_{ox} is the oxide thickness, and a_l, a_v, and a_w are process-dependent coefficients. Typical values for these coefficients are 7, 1, and 10, respectively. It is easy to see that the above expression also models an effective reduction of the body effect for short-channel devices, and an effective increase of the same effect for narrow-channel devices. The presence of V_{DS} in Eq. (3.21) makes necessary the calculation of a new value of V_{DS}', obtained as the value of V_{DS} at which the nonsaturation slope of the $I_D - V_{DS}$ curve becomes equal to the saturation slope [1, 2]. From Eq. (3.21) it becomes clear that as far as the absence of short- and narrow-channel effects for a given fabrication process is concerned, what counts is not the size of L and W per se, but rather how large they are in comparison to the oxide thickness.

6. When one is designing fabrication processes for very small devices, the series resistance R_S of the source can drop a nonnegligible voltage $I_D R_S$; then the effective gate-source voltage applied to the device is only $V_{GS,eff} = V_{GS} - I_D R_S$. Thus I_D becomes smaller, and increases with V_{GS} less rapidly, than what our simple models would predict. The effect of the transistor characteristics is qualitatively the same as that of the dependence of mobility on bias, which we discussed in Sec. 3.6. In fact, the two effects are sometimes lumped together and modeled by an appropriate expression for μ as a decreasing function of V_{GS}. This is misleading, since R_S has nothing to do with the mobility in the channel. A more straightforward approach is to model R_S with a separate resistor external to the device.

7. A phenomenon known as *punch-through* occurs in devices with very short channels. As V_{SB} and/or V_{DS} is increased, the depletion regions surrounding the source and drain can spread enough to approach each other; a large current known as the *punch-through current* can then flow below the surface. Ion implantation can be used to reduce punch-through effects, by increasing the doping concentration at such points and thus keeping the depletion region widths small.

8. If L is very small, and thus the voltage V_{DS} must be dropped over a very short distance, then the resulting electric field parallel to the surface can be high. The electrons can then reach a maximum velocity, called the *saturation velocity*, corresponding to a limited current. The

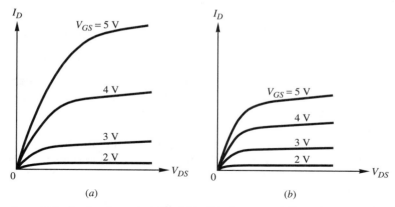

Figure 3.15 Device characteristics (a) in the absence and (b) in the presence of velocity saturation effects.

I_D-V_{DS} curves can then saturate "early," before they can reach the values of V_{DS} and I_D that would be predicted by the model in Table 3.1. This is illustrated in Fig. 3.15, in which we show characteristics calculated under the assumption that velocity saturation is absent (Fig. 3.15a) and present (Fig. 3.15b). The spacing of the curves in the presence of velocity saturation can become nearly constant for equal V_{GS} increments, as shown in Fig. 3.15b; this is to be contrasted with the square-law spacing in Fig. 3.15a. In addition, since in saturation the electrons enter the drain region at maximum velocity, the amount of charge transferred by them per unit of time (i.e., the drain current) is fixed by V_{GS} and W and does not depend on L. This is just like water coming out of a pipe at constant speed: The amount of water running per unit time is independent of the length of the pipe. In an effort to save the "long-channel" models in the presence of velocity saturation, sometimes expressions such as those of Table 3.1 are used in conjunction with an "effective" mobility which now is made a function of V_{DS}; taking into account that this parameter is also made a function of V_{GS} and V_{SB} (Sec. 3.6), we see that the result is a complicated expression for effective mobility. One usually avoids the use of such complicated expressions in hand analysis and relies on computer simulation for taking such effects into account. For approximate results if V_{DS}/L is not too large [33, 1–4], one can use

$$\mu = \frac{\mu'}{1 + V_{DS}/(LE_c)} \qquad (3.22)$$

where μ' is the effective mobility in the absence of short-channel effects (but ideally including the effect of V_{GS} and V_{SB}, as in Sec. 3.6) and E_c is a critical electric field value (typically 2 V/μm). As in the case of Eq. (3.21), the presence of V_{DS} makes necessary a new value of V'_{DS}, obtained as the value of V_{DS} at which the nonsaturation slope of the $I_D - V_{DS}$ curve becomes equal to the saturation slope [1]. Such complicated modifications are, of course, only feasible in computer-aided analysis and are included in most CAD models for the MOSFET.

9. The high electric fields present in short-channel devices can impart high energy on electrons and holes. These carriers are then said to be *hot;* they can cause breakdown problems (Sec. 3.9), can generate significant substrate currents, and can even enter the oxide where they can be trapped, gradually accumulating over time and degrading device performance (e.g., causing a change in V_{T0}). Hot carrier effects must be kept low through appropriate device design techniques and through limiting V_{DS}.

10. We have already described the effect of channel length modulation in the context of long-channel devices. From the discussion in Sec. 2.4.3, or from Eq. (3.3) and the formula for V_A in Table 3.3, one expects this effect to become more pronounced for shorter channels, and indeed this is the case.

Scaling. The above small-dimension effects can obviously become severe if one attempts to fabricate devices smaller than those for which a process was intended. To a large extent, these effects can be reduced if the fabrication process and device bias are appropriately modified. To get a feel for the considerations involved, recall first that L cannot be made smaller than about the sum of the depletion regions around the source and around the drain, since when these two regions approach each other, punch-through can occur (see above); to be able to make L small, then, the width of these regions must be small. This can be achieved by reducing their reverse bias and increasing the body doping concentration. To make sure that the reverse bias remains small under any circumstances, the power supply voltage must be reduced. The increase in body doping will cause an increase in threshold voltage (Sec. 2.8). This will make it more difficult to turn the devices sufficiently "on," especially since the power supply voltage has been reduced; to lower the threshold voltage, the oxide thickness must be reduced. The threshold voltage value is also "tailored" through ion implantation.

Process modifications in the above spirit are known as *scaling.* Several scaling scenarios have been proposed to reduce device dimensions; these are summarized and compared elsewhere [1]. In general, the trend is toward smaller horizontal dimensions, smaller oxide thickness, smaller junction depth, heavier substrate doping, and (although not always) lower power supply voltage.

3.9 Breakdown

The value of the various voltages that can be applied to a MOS transistor should be limited, in order to avoid several forms of breakdown. One such form is *junction breakdown:* The junctions formed by the substrate and the drain or source regions will conduct a large current if the reverse bias applied to them exceeds a certain value. This form of breakdown will occur even with the device off.

When the device is on, carriers moving fast in the channel can impact on silicon atoms and ionize them, producing electron-hole pairs. These in turn can speed up, impact on other atoms, and produce more electron-hole pairs, etc. Currents larger than those predicted by the normal current equations will then flow. This phenomenon is referred to as *channel breakdown.*

As already discussed in Sec. 3.8, *punch-through* will occur when the depletion regions surrounding source and drain approach each other; this will be especially noticeable in devices with very short channels, and it can be considered another form of breakdown.

The above breakdown mechanisms are nondestructive; once the large voltages producing them are removed, the device will function as long as no permanent damage due to overheating has occurred (possibly with some degradation in performance due to hot electron effects; see Sec. 3.8, item 9). The effect of such breakdown on device characteristics is shown in Fig. 3.16.

A *destructive* breakdown mechanism is *oxide breakdown;* it occurs when the electric field in the gate insulator exceeds a certain value (about 6 V per 100 Å in silicon dioxide). The result is a permanent short circuit through the insulator. Static charge, such as that transferred to gates by handling devices with bare hands, is known to cause oxide breakdown. For this reason, in integrated circuits protective devices

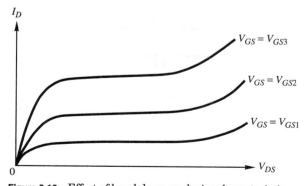

Figure 3.16 Effect of breakdown on device characteristics.

are often used at those input terminals that are connected to transistor gates (Chap. 5).

Device engineers establish safe operating limits in terms of bias voltages which ensure that no form of breakdown occurs. These limits often take the form of a single specified maximum allowed power supply voltage, with the understanding that no other signal applied to the integrated circuit can exceed the potential of either power supply rail; a common power supply voltage value for several years has been 5 V. Lower supply voltages (e.g., 3.3 or 2.5 V) are increasingly used in conjunction with modern fabrication processes and are attractive in battery-operated equipment (1.5 V is used for watch circuits). Circuits operating with supply voltages as low as 0.9 V are sometimes used.

3.10   The *p*MOS Transistor

If the substrate is made of n-type material, and the source/drain regions are made of p-type material, we have what is known as the

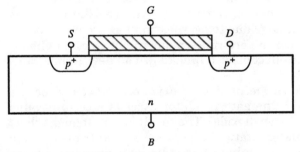

Figure 3.17   A *p*MOS transistor.

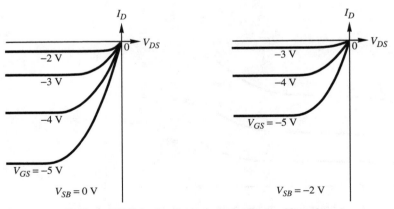

Figure 3.18   Typical *p*MOS device characteristics for two V_{SB} values.

p-channel MOS transistor, or *pMOS transistor* for short. Figure 3.17 shows such a device. An example of *p*MOS transistor characteristics is shown in Fig. 3.18.

The operation of the *p*MOS transistor is the dual of *n*MOS operation. The role of electrons is played by holes, and the role of ionized acceptor atoms is played by ionized donor atoms. Statements made about *n*MOS devices can be adapted to the case of *p*MOS devices with simple modifications. For example, in Fig. 3.17 the more *negative* the drain-source voltage, the heavier the flow of *holes* from source to drain, hence, the more *negative* the drain current (assuming the reference direction is chosen as before, i.e., from the drain through the channel to the source). Also, the more *negative* the gate-source voltage, the heavier the concentration of *holes* in the channel and the more *negative* the drain current. The zero-substrate-bias threshold voltage V_{T0} is *negative* for enhancement-mode and *positive* for depletion-mode *p*MOS devices.

The above qualitative duality aside, there exists an important quantitative difference between *n*MOS and *p*MOS devices: In the latter, the effective mobility is much smaller (about 3 times). Thus a typical value for μ in *p*MOS devices is 200 cm²/(V · s).

Parameters I'_M, V_A, ϕ_0, and ϕ_α (Tables 3.2 and 3.3) are, in principle, negative quantities for *p*MOS devices, but are often specified as positive quantities by convention. We will use absolute value signs when specifying these parameters for *p*MOS devices, to eliminate ambiguity.

The model equations for *p*MOS devices can be obtained from *n*MOS device equations with rather obvious sign changes or the appropriate use of absolute values. For example, consider the body effect equation, giving the variation of V_T with V_{SB}. In *p*MOS devices, the more *negative* the value of V_{SB}, the more *negative* (or less positive) V_T will be; thus

$$V_T = V_{T0} - \gamma(\sqrt{|V_{SB}| + |\phi_0|} - \sqrt{|\phi_0|}) \qquad (3.23)$$

The equations for the drain current can be converted in a similar manner (Prob. 3.27).

One must exercise extra care in view of conflicting conventions for the sign of some quantities mentioned above. Whenever in doubt, one can follow a common engineering practice that reduces the chance of sign errors: To determine current or voltage quantities for a *p*MOS device, first determine such quantities for a fictitious, *corresponding* *n*MOS device, and then reverse the sign of the result if necessary. This is illustrated in the following example.

Example 3.11 A *p*MOS device with $W = 10$ μm, $L = 2$ μm, $K' = 28$ μA/V², $V_{T0} = -0.9$ V, $\gamma = 0.7$ V$^{1/2}$, $|\phi_0| = 0.72$ V, and $|\phi_\alpha| = 1$ V is biased with $V_{GS} = -3$ V, $V_{SB} = -1$ V, and $V_{DS} = -2$ V. Find V_T, V'_{DS}, and I_D.

solution   The corresponding fictitious nMOS device has parameter values of the same magnitude as those of the given pMOS device, the same W and L, and bias voltages $V_{GS} = +3$ V, $V_{SB} = +1$ V, and $V_{DS} = +2$ V. Using Tables 3.1 and 3.2, we find $V_T = 1.22$ V, $V'_{DS} = 1.47$ V, and $I_D = 183$ μA. Thus, for the given pMOS device we have

$$V_T = -1.22 \text{ V}$$

$$V'_{DS} = -1.47 \text{ V}$$

$$I_D = -183 \text{ μA}$$

Note that just as the fictitious nMOS device is in saturation since V_{DS} is more positive than its $V'_{DS} = 1.47$ V, the given pMOS device is in saturation since V_{DS} is more negative than its $V'_{DS} = -1.47$ V.

3.11   Device Symbols

Common enhancement-mode device symbols are shown in Fig. 3.19. Symbols a and d, or c and f emphasize source-drain symmetry; b and e do not. In all cases, the body terminal can be omitted if it is understood from the context where it is connected. In the case of depletion-mode devices, similar symbols are used except the "channel" is usually "fat-

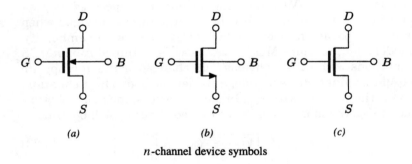

(a)   (b)   (c)

n-channel device symbols

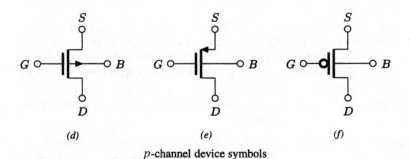

(d)   (e)   (f)

p-channel device symbols

Figure 3.19   Common symbols for enhancement-mode MOS transistors.

Figure 3.20 Common symbols for depletion-mode
nMOS transistors.

tened" as illustrated, for example, in Fig. 3.20. Other symbols are also
in use for both enhancement- and depletion-mode devices.

3.12 Model Accuracy, Parameter Extraction, and Computer Simulation [1]

Matching a model to measurements. The models we have presented are
based on simple theories, which predict certain values for the parameters involved. For example, the equation for γ in Table 3.3 gives the
value of the body effect coefficient in terms of substrate concentration
and oxide capacitance per unit area. As already mentioned in Sec. 3.2,
such formulas are exact only for fictitious, idealized devices corresponding to simplifying assumptions which are made in the course of
model development [1]. There are numerous simplifying assumptions
behind any simple model, and thus the resulting model equations cannot be expected to perfectly describe a real device; such a device, for
example, does not have an exactly uniform substrate. Thus, to help the
"ideal" equations model a *real* device, it is common to choose values for
the model parameters in such a way that good matching between
model predictions and experimental measurements is achieved [2, 3].
In this *parameter extraction* process, the "theoretical" values for model
parameters, such as given in Table 3.3 for γ, might simply play the role
of an initial guess, which then might have to be modified to achieve
best matching.

Of course, what constitutes "best matching" is subject to interpretation. For digital circuits, for example, one might desire good matching
for devices with small channel lengths and $V_{SB} = 0$, operating in strong
inversion or cutoff; then the model parameters will be optimized for
matching in the restricted channel length and bias ranges encountered
in this application. If the model is used outside these ranges, the matching may not be satisfactory. For example, it is unlikely that the parameters so chosen will be satisfactory for analog circuits, which employ
devices of various channel lengths and various V_{SB} values, operating
anywhere from weak to strong inversion. To make possible acceptable
matching to measurements in such cases, new values may have to be
assigned to some of (or all) the model parameters. Since these new val-

ues represent a compromise to make matching acceptable over wide ranges, it can be expected that the degree of matching achieved in this case will be worse than in the restricted ranges associated with the digital application mentioned earlier. If the resulting matching is not acceptable, sometimes "binning" is used; for example, the expected range of L values is divided into several subranges, and different model parameter values are used within each subrange. This, of course, has the drawback that if L is near the boundary between two subregions, a slight change in its value can result in the model's artificially predicting an abrupt change in the device and circuit performance.

To avoid the above problems, one may have to use a better model which is likely to be more complicated and hence computationally slower. In computer simulation programs, one is often provided with models of various "levels," each representing a different compromise between accuracy and simplicity. Note that if the same parameter (say, γ) appears in two different models, different values of it may have to be used in each model for best results. For this reason, direct comparison of the performance of different models is difficult, unless it is known that parameter values have been separately optimized for each model.

Let us offer an example of problems that can arise when model predictions are compared to experimental results. Let us say that the weak inversion model of Table 3.1 for given terminal voltages and process parameters predicts a current 4 times larger than the measured value. Is the model bad? Not necessarily; blind number comparison can be extremely misleading. The problem may just be that the process parameters are not accurately known. With $n = 1.3$, just a 50-mV horizontal shift in Fig. 2.12b is enough to give 4 times the actual current! This, of course, happens because in weak inversion the $\log I_D$ versus V_{GS} curves are so steep that a slight shift horizontally corresponds to a large current change. Note that if I-V plots were compared rather than numbers, the above problem could have been spotted immediately. One would have caught the fact that the curves produced by the model had the same shape as the measured ones, except that they were shifted somewhat horizontally. Thus V_M would be suspected, and a new value would be tried for that parameter.

Parameter extraction.   Let us now give a simple example of parameter extraction for the approximate strong inversion model of Table 3.1. To determine K and V_T empirically for a given device, note that, at very small V_{DS}, we obtain

$$I_D \simeq KV_{DS}(V_{GS} - V_T), \qquad \text{very small } V_{DS} \qquad (3.24)$$

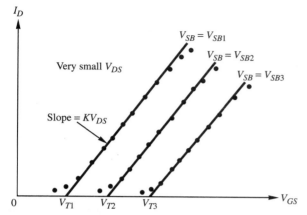

Figure 3.21 Graphical method for determining K and V_T from strong inversion nonsaturation measurements with very small V_{DS}.

Values of I_D and V_{GS}, as obtained from measurements for very small V_{DS} and for fixed values of V_{SB}, can be plotted as shown in Fig. 3.21. Straight lines are fitted through the points. The deviation from straight-line behavior at low and high V_{GS} values is due to moderate inversion and effective mobility degradation, respectively.* The straight-line parts are extended to the horizontal axis. The slope of these lines gives KV_{DS}, as seen from Eq. (3.24); from this and the known, fixed V_{DS}, K can be found. If, in addition, W/L is known, K' can be found from $K = (W/L)K'$. If the slope is not exactly the same for all curves, an average should be taken.

The intercept of each curve gives V_T for each V_{SB} value, as seen from Eq. (3.24). (If V_{DS} is not completely negligible, a small correction will be needed; see Prob. 3.31.) The V_T values thus obtained can be plotted versus $\sqrt{V_{SB} + \phi_0} - \sqrt{\phi_0}$, as shown in Fig. 3.22, assuming a value for ϕ_0, say, 0.7 V. If these points fall approximately on a straight line, they can be modeled by the equation for V_T in Table 3.2; otherwise, a different value for ϕ_0 can be tried until a straight-line fit is satisfactory. The slope of the straight line is then the body effect coefficient γ; the intercept is V_{T0}. Notice that the value of V_{T0} found in this way might be a little off from the "exact" value that can be found from Fig. 3.21 with $V_{SB} = 0$; this is a price we might have to pay for an overall satisfactory fit in Fig. 3.22.

---

* Recall that K is given by $(W/L)\mu C'_{ox}$ (see Tables 3.1 and 3.3). The effective mobility is assumed to drop at high V_{GS} values, as explained in Sec. 3.6. The parameter μ will be assumed here to be the value of the effective mobility at *low* V_{GS} values, before the effect of V_{GS} on the effective mobility becomes noticeable.

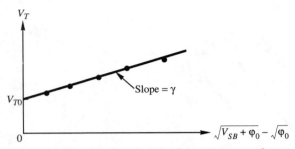

Figure 3.22   Graphical method for determining γ and ϕ_0.

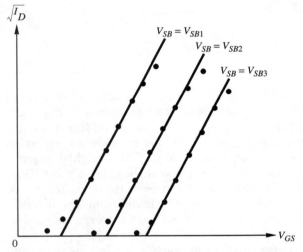

Figure 3.23   Graphical method for determining V_T and k from strong inversion saturation measurements.

An alternative procedure for finding V_T would be to plot $\sqrt{I_D}$ versus V_{GS} in saturation, as shown in Fig. 3.23. From Tables 3.1 and 3.2, the result should be a straight line with an intercept of V_T and a slope of $\sqrt{K/(2\alpha)}\,(V_{GS} - V_T)$. Note that the values of V_T determined in Fig. 3.23 are those for best fit in the *saturation* region; they might not be exactly the same as those found in Fig. 3.21.*

---

* Similarly, the value of ϕ_0 found as explained above will not be exactly the same as that found through other techniques using weak inversion measurements. In fact, it can be shown that normally the value of ϕ_0 to be used in the V_T equation should be somewhat larger than the value used in the equation for the weak inversion parameter n (Table 3.2). The two ϕ_0's encountered in these two equations are actually different parameters [1], although for simplicity we adopted the common simplification to consider them as the same quantity.

We take the opportunity here to emphasize that in our discussion V_T is the *extrapolated* threshold voltage, a term we first met in Sec. 2.7 and which is justified by the construction in Fig. 3.21. Unfortunately, the term *threshold voltage* is used in the literature with at least two other meanings. Sometimes it is used for what is more properly called "constant current threshold voltage." This is the value of V_{GS} needed to reach a set value of $I_D/(W/L)$, which often happens to fall somewhere in the moderate inversion region or even in the weak inversion region. Elsewhere, threshold voltage is taken to mean vaguely "the value of V_{GS} at which inversion begins," without specifying which level of inversion is meant (often strong inversion is implied). The term "threshold voltage" is often used indiscriminately for these quantities, and sometimes during parameter extraction the attempt is made to match calculated values of one of these "thresholds" to measured values of another threshold (e.g., the extrapolated threshold). All this originated in the early days, when V_{GS} values of 20 V were not uncommon; compared to these, an ambiguity of, say, 0.3 V in the value of the threshold voltage was not a problem. Today, however, when the trend is toward low-voltage circuit operation, a distinction in the above quantities is important. In this book "threshold voltage" will always imply *extrapolated* threshold voltage.

The determination of parameter values, which we have illustrated above by means of a simple example and with the help of Figs. 3.21 to

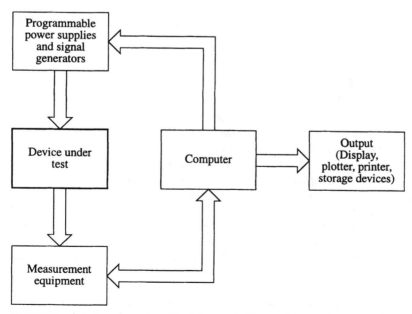

Figure 3.24 A system for automatic data acquisition and parameter extraction.

3.23, is done in a much more sophisticated way [2, 3] in automatic systems devoted to data acquisition and parameter extraction. These take the general form shown in Fig. 3.24. Under the central control of a computer, measurements such as $I - V$ characteristics are taken on many devices of a certain type. The data obtained are then used as input to software which contains algorithms for the determination of parameter values. Crucial in this process is the minimization of a certain error (e.g., mean square error) between the measured data and the values predicted by the model used.

Distortion modeling. An application which requires accurate large-signal modeling is the prediction of distortion (deviation from the sinusoidal response that would be expected from a linear circuit). The distortion value can be strongly dependent on fine nuances in the transistor characteristics, especially when it is small. This is illustrated in Fig. 3.25 [34], which shows a measured curve in a and a prediction from a model in b; it is seen that the model provides excellent accuracy for I_D. If now a sinusoidal signal is added to the voltage, the corresponding current *signal* predicted by the model is very different from the one seen experimentally, due to differences in the *fine nuances* of the two curves! Fourier analysis of the measured and predicted current signals will result in totally different distortion values. The conclusion that can be reached from the above qualitative discussion is clear. In cases where the distortion is small, in general one should not trust computer simulation. A designer may have to go to extra pains to obtain a realistic value of the distortion expected from a given circuit.

Models in computer simulators. Numerous models used for computer simulation are discussed elsewhere [1–4, 6–19, 26, 28, 31]. There is currently significant activity in testing various models, and this should eventually result in the wide adoption of the better ones among them. The reader is warned that some models currently in use in computer simulators contain errors and are especially inadequate for analog and analog/digital design. Nevertheless, blind trust in the computer is common, especially among novice designers. Eventually many learn the hard way not to trust the computer, i.e., after they get to test their chips and find out that the chips do not meet the specifications. To remedy these problems, one can use better models, or can at least become aware of the limitations of the models one uses, and then design conservatively (a "just in case" approach). The many high-performance analog MOS circuits already in use are living proof that good circuits can be designed, despite the present poor state of computer simulator MOSFET models for analog design. Such circuits,

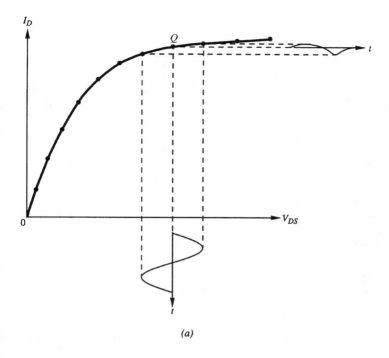

(a)

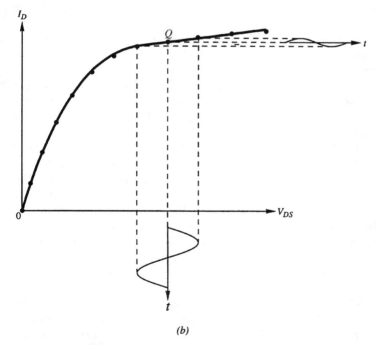

(b)

Figure 3.25 The I_D-V_{DS} curve as (*a*) measured and (*b*) modeled. For a sinusoidal voltage variation, the shape of the corresponding current waveform in (*b*) is severely in error, although the error in modeling I_D is very small throughout. (*Copyright 1994 by IEEE.*)

however, often come at the expense of extra pains during the design stage, or they are designed so conservatively that potential performance is sacrificed. So the need for better models in simulators is still pressing. Note that this does not always imply new work in modeling per se; there are certain good modeling ideas that "sit" in research journals for many years before (if ever) they find their way into simulators. For example, simple nonquasistatic models (Chap. 4) were described in the literature in the 1960s; yet many popular simulators still lack such models. In other cases good models have been implemented in computer simulators, but either they are proprietary or, if openly available, there is a lot of inertia in the technical community to accepting them.

We have seen that by empirically "adjusting" the parameter values of a model one can extend its region of validity or reduce the error in certain key quantities. *If the model is physically based, the final parameter values one ends up with after extraction will be close to the theoretical values.*

CAD models are usually complex since they have to take into account carefully several physical phenomena which the simpler models ignore. By the time all regions of operation are included, one can easily end up with nightmarish models with, say, 40 parameters. These can provide accuracy at the expense of complexity.

On the other hand, satisfactory accuracy and a large number of parameters do not always mean that the physics of the device has been modeled correctly. There exist models which are based on faulty premises but which, just because they contain many parameters, can give adequate accuracy after extensive empirical adjustment. A serious drawback of such models is that they mislead the user. Consider, for example, a strong inversion model used in moderate inversion (a practice seen often). In an attempt to "fit the unfittable," a parameter extraction system will try all kinds of variations on the parameter values, and it might end up, for example, with a fictitious value for the oxide thickness. The complete current equation now might fit the experimental data reasonably well, but the artificial value forced on the oxide thickness may mislead the user and will cause an error in predicting the capacitances of the device. (Our example concerning the oxide thickness may be fictitious, but similar cases are not rare.) Another major problem with models based on wrong physical assumptions is that they are likely to fail in *predicting* what will happen if some fabrication process parameters are changed in the future. Such predictions are invaluable, yet experimental data are not available at the time the prediction is needed and thus curve fitting cannot be done. A good model based on correct physical assumptions is then the only resort.

At the other extreme from models with too many parameters to "fiddle with" are models which just do not have enough parameters to adjust and which fail to match experimental results no matter how much curve fitting is attempted. An example is the widely used strong inversion equations which implicitly assume that our parameter α (Tables 3.1 and 3.2) is equal to 1; this has been discussed in Example 3.5.

Developing a model is an art involving constant tradeoffs between accuracy and simplicity. Again, we emphasize that models based on correct physical assumptions are invaluable, and that for such models, the parameter values one ends up with after parameter extraction will be close to the values predicted by theory. Circuit designers can benefit from such models. At any rate, they must at least be aware of the capabilities and limitations of the models they are using. Benchmark tests [34] for the evaluation of models used in a computer simulator can be found in App. B. These can be better appreciated after one has studied Chap. 4.

References

1. Y. P. Tsividis, *Operation and Modeling of the MOS Transistor,* McGraw-Hill, New York, 1987.
2. H. C. de Graaff and F. M. Klaassen, *Compact Transistor Modeling for Circuit Design,* Springer-Verlag, New York, 1990.
3. N. Arora, *MOSFET Models for VLSI Circuit Simulation—Theory and Practice,* Springer-Verlag, New York, 1993.
4. K. Lee, M. Shur, T. A. Fjeldly, and T. Ytterdal, *Semiconductor Device Modeling for VLSI,* Prentice-Hall, Englewood Cliffs, NJ, 1993.
5. R. J. Van Overstraeten, G. J. Declerck, and P. A. Muls, "Theory of MOS Transistor in Weak Inversion—New Method to Determine the Number of Surface States," *IEEE Transactions on Electron Devices,* 22:282–288, May 1975.
6. G. Merckel, J. Borel, and N. Z. Cupcea, "An Accurate Large-Signal MOS Transistor Model for Use in Computer-Aided Design," *IEEE Transactions on Electron Devices,* 19:681–690, May 1972.
7. F. M. Klaassen, "MOS Device Modeling," in Y. Tsividis and P. Antognetti, eds., *Design of VLSI Circuits for Telecommunications,* Prentice-Hall, Englewood Cliffs, NJ, 1985.
8. L. W. Nagel, *SPICE2: A Computer Program to Simulate Semiconductor Circuits,* Rep. ERL-M520, Electronics Research Laboratory, University of California at Berkeley, 1975.
9. P. Antognetti and G. Massobrio, *Semiconductor Device Modeling with Spice,* McGraw-Hill, New York, 1993.
10. D. Divekar, *FET Modeling for Circuits Simulation,* Kluwer, Boston, 1988.
11. G. T. Wright, "Physical and CAD Models for the Implanted-Channel VLSI MOSFET," *IEEE Transactions on Electron Devices,* 34:823–833, April 1987.
12. A. R. Boothroyd, S. W. Tarasewicz, and C. Slaby, "MISNAN—A Physically Based Continuous MOSFET Model for CAD Applications," *IEEE Transactions on Computer-Aided Design,* 10:1512–1529, December 1991.
13. J. A. Power and W. A. Lane, "An Enhanced Spice MOSFET Model for Analog Applications," *IEEE Transactions on Computer-Aided Design,* 11:1418–1425, November 1992.
14. N. D. Arora, R. Rios, C.-L. Huang, and K. Raol, "PCIM: A Physically Based Continuous Short-Channel IGFET Model for Circuit Simulation," *IEEE Transactions on Electron Devices,* 41:988–997, June 1994.

15. J. H. Huang, Z. H. Liu, M. C. Jeng, P. K. Ko, and C. Hu, "A Physical Model for MOS-FET Output Resistance," *Digest of Technical Papers, International Electron Devices Meeting,* San Francisco, CA, 1992.
16. J. H. Huang, Z. H. Liu, M. C. Jeng, P. K. Ko, and C. Hu, "A Robust Physical and Predictive Model for Deep-Submicron MOS Circuit Simulation," *Proceedings of the IEEE Custom Integrated Circuits Conference,* San Diego, CA, May 1993, pp. 14.2.1–14.2.4.
17. S. M. Gowda, B. J. Sheu, and C.-H. Chang, "Advanced VLSI Circuit Simulation Using the BSIM-Plus Model," *Proceedings of the IEEE Custom Integrated Circuits Conference,* San Diego, CA, May 1993, pp. 14.3.1–14.3.5.
18. R. M. D. A. Velghe, D. B. M. Klaassen, and F. M. Klaassen, "Compact MOS Modeling for Analog Circuit Simulation," *Digest of Technical Papers, International Electron Devices Meeting,* Washington, DC, 1993, pp. 485–488.
19. T. Skotnicki, C. Denat, P. Senn, G. Merckel, and B. Hennien, "MASTAR: A New Analog/Digital CAD Model for Sub-halfmicron MOSFETs," *Proceedings of the International Electron Devices Meeting,* San Francisco, CA, 1994.
20. H. K. J. Ihantola and J. L. Moll, "Design Theory of a Surface Field-Effect Transistor," *Solid State Electronics,* 7:423–430, June 1964.
21. J. A. Van Nielen and O. W. Memelink, "The Influence of the Substrate upon the D.C. Characteristics of Silicon MOS Transistors," *Philips Research Reports,* 22:55–71, February 1967.
22. K. A. Valiev, A. N. Karmazinksii, and A. M. Korolev, *Digital Integrated Circuits Using MOS Transistors,* Soviet Radio, Moscow, 1971.
23. E. Vittoz and J. Fellrath, "CMOS Analog Integrated Circuits Based on Weak Inversion Operation," *IEEE Journal of Solid State Circuits,* 12:224–231, June 1977.
24. H. Oguey and S. Cserveny, "MOS Modeling at Low Current Density," Summer Course on Process and Device Modeling, K U Lueven, 1983.
25. E. Vittoz, "Micropower Techniques," in J. E. Franca and Y. Tsividis, eds., *Design of Analog-Digital VLSI Circuits for Telecommunications and Signal Processing,* Prentice-Hall, Englewood Cliffs, NJ, 1994.
26. C. Enz, "High Precision CMOS Micropower Amplifiers," Ph.D. thesis no. 802, Ecole Polytechnique Federal de Lausanne, Switzerland, 1989.
27. E. A. Vittoz, "Low-Power Design; Ways to Approach the Limits," *Digest of Technical Papers, 1994 IEEE International Solid-State Circuits Conference,* 1994, pp. 14–18.
28. C. C. Enz, F. Krummenacher, and E. A. Vittoz, "An Analytical MOS Transistor Model Valid in All Regions of Operation and Dedicated to Low-Voltage and Low-Current Applications," *Analog Integrated Circuits and Signal Processing,* 8:83–114, July 1995.
29. Y. Tsividis, "Moderate Inversion in MOS Devices," *Solid-State Electronics,* 25:1099–1104, 1982. See also Erratum, ibid., 26:823, 1983.
30. J. R. Brews, "A Charge Sheet Model for the MOSFET," *Solid-State Electronics,* 21:345–355, 1978.
31. C. McAndrew, B. K. Bhattacharyya, and O. Wing, "A Single-Piece C_{inf} Continuous MOSFET Model Including Subthreshold Conduction," *IEEE Electron Device Letters,* 12:565–567, 1991.
32. Y. Tsividis, K. Suyama, and K. Vavelidis, "A Simple 'Reconciliation' MOSFET Model Valid in All Regions," *Electronics Letters,* 31: no. 6, 506–508, March 1995.
33. F. N. Trofimenkoff, "Field-Dependent Analysis of the Field Effect Transistor," *Proceedings of the IEEE,* 53:1765–1766, January 1965.
34. Y. Tsividis and K. Suyama, "MOSFET Modeling for Analog Circuit CAD: Problems and Prospects," *IEEE Journal of Solid State Circuits,* 29:210–216, March 1994.

Problems

Note: In all problems, the transistors used are assumed to be *n*MOS devices at room temperature (300 K), unless noted otherwise. Mobility dependence on bias, and short- and narrow-channel effects are to be neglected unless stated otherwise.

3.1 Consider a transistor operating in strong inversion saturation. It has been claimed that a way to eliminate the body effect is to apply to the device a gate-source bias of the form $V_{GS} = V_C + V_T$, where V_C is constant and V_T is equal to the threshold voltage of the device (there are circuits that can produce such a bias voltage). Explain why this approach will not eliminate the dependence of I_D on V_{SB}.

3.2 Two identical devices are operating in weak inversion saturation with the same V_{SB}. Find an expression for the ratio of their drain currents, and show that I_M and V_M are not needed to determine this ratio.

3.3 For an nMOS device with electrical parameters as given in Example 3.8, $W = 4$ μm, and $L = 6$ μm, plot I_D versus V_{DS}, with V_{GS} as a parameter, for the following two cases:

(a) $V_{SB} = 0$ V and $V_{GS} = 2, 3, 4$, and 5 V
(b) $V_{SB} = 2$ V and $V_{GS} = 2$ and 3 V

To avoid entering the breakdown region, $V_{DB} = V_{DS} + V_{SB}$ should not exceed 5 V.

3.4 For the device of Prob. 3.3 plot I_D versus V_{DS} in weak inversion, with V_{GS} a parameter, assuming $V_M = V_T - 2\phi_t$. Use values of V_{DS} between 0 and 1 V and V_{GS} such that $V_{GS} - V_M = -0.05, -0.10, -0.15$, and -0.2 V. Give one plot for $V_{SB} = 0$ V and another for $V_{SB} = 2$ V.

3.5 For the device of Probs. 3.3 and 3.4, plot the weak inversion I_D (on a logarithmic axis) versus V_{GS} in saturation, for $V_{SB} = 0, 1$, and 2 V.

3.6 Find the region of inversion in which a device operates, assuming that the electrical parameters are as in Example 3.8, $V_{DS} = 5$ V, and $V_{SB} = 0$ V, in the following cases:

(a) $W/L = 0.1$, $I_D = 1$ μA
(b) $W/L = 200$, $I_D = 1$ μA
(c) $V_{GS} = 1.5$ V
(d) $V_{GS} = 0.65$ V

3.7 Find the saturation I_D value required to bring a device to (a) the upper limit of weak inversion and (b) the lower limit of strong inversion, assuming $W = 60$ μm, $L = 3$ μm, $V_{SB} = 0$ V, and the electrical parameters as in Example 3.8.

3.8 Plot V_T, α, and k' versus V_{SB} for the device of Example 3.8. Use V_{SB} values from 0 to 5 V.

3.9 Find V_A if it is known that in saturation I_D is 203 μA at $V_{DS} = 3$ V and is 209 μA at $V_{DS} = 4$ V. If the parameters of the device are as given in Example 3.8, estimate the channel length of the device.

3.10 Show that at very low V_{DS} values the MOS transistor in strong inversion can be considered an approximately linear resistor. Find the expression for its resistance, and show that it can be controlled through V_{GS}. What condition must be imposed on V_{DS}, if it is desired that the *I-V* characteristic of this "resistor" not deviate from linearity by more than 10 percent? 1 percent? 0.1 percent?

3.11 (*a*) A device in weak inversion saturation has $I_D = 29$ nA at $V_{GS} = 0.63$ V and $I_D = 87$ nA at $V_{GS} = 0.67$ V. Find the value of n and the value of $I_{DC} = I_M \exp[-V_M/(n\phi_t)]$.
(*b*) If in addition it is given that $V_M = 0.7$ V and $W/L = 10$, find I_M and I'_M.

3.12 Consider an *n*MOS device with characteristic curves given in Figs. 2.9 and 2.10. From these curves, estimate the values of parameters n, k, V_{T0}, α, and γ for this device. Assume $\phi_0 = 0.7$ V. Also, determine approximate values for I_M, V_M, I_H, and V_H at $V_{SB} = 0$ V and $V_{SB} = 2$ V.

3.13 Suggest a graphical technique for determining the value of n in weak inversion.

3.14 A device characterized by the parameters given in Example 3.8 has $W/L = 10$. With $V_{GS} = 2$ V and $V_{DS} = 4$ V, it gives $I_D = 377$ μA. What is the value of V_{SB}? Solve this problem by using an iterative approach as follows: Assume α is approximately equal to its value at $V_{SB} = 0$; find V_T, and from this find V_{SB}. Find the corresponding value of α, recalculate V_T, and from this find V_{SB} again. Note that this second iteration gives practically the exact value of V_{SB}, and that even a single iteration would have sufficed for some applications. Approximate iterative solutions, rather than exact ones, are often used in circuit work.

3.15 Calculate the quantities indicated in Fig. 3.26, assuming that V_A is infinite and all other parameters are as given in Example 3.8, unless noted otherwise. The quantities next to the devices in the form a/b mean $W = a$ and $L = b$. Neglect short- and narrow-channel effects.

3.16 Calculate the quantities indicated in Fig. 3.27, assuming that all parameters are as given in Example 3.8, unless noted otherwise. The quantities next to the devices in the form a/b mean $W = a$ and $L = b$. Make reasonable approximations where warranted. (*Warning:* One of the cases is meant to be tricky.)

3.17 Develop an expression for I_Y versus V_{YX} in Fig. 3.7.

3.18 Show that the strong inversion nonsaturation symmetric model of Eq. (3.5) predicts $dI_D/dV_{DB} = 0$ at $V_{DB} = V_P$, where V_P is given by Eq. (3.6). Extend the model for use in the saturation region. Use an approach similar to that in Table 3.1.

3.19 Show that the strong inversion model in Table 3.1 can be derived from the symmetric model in Eq. (3.5). (*Hint:* Express all voltages in terms of V_{GS}, V_{SB}, and V_{DS}; use a series expansion for the first ¾-power term around $V_{DS} = 0$; and keep the first three terms of the series.) Show that the accuracy of the approximation can be improved if the coefficient of the second-order term in the series is modified.

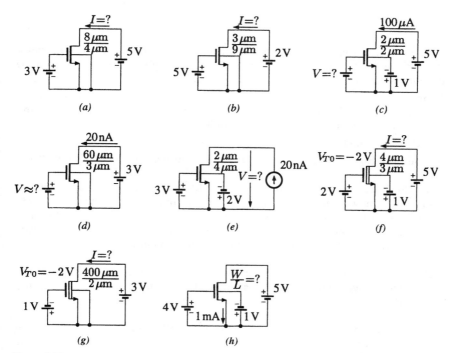

(a)　(b)　(c)

(d)　(e)　(f)

(g)　(h)

Figure 3.26

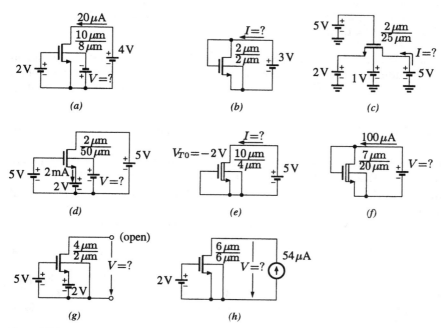

(a)　(b)　(c)

(d)　(e)　(f)

(g)　(h)

Figure 3.27

3.20 (*a*) Using appropriate series expansions, show that for low-voltage work the symmetric strong inversion nonsaturation model in Eq. (3.5) can be approximated by the model of Eqs. (3.7) and (3.8).

(*b*) Show that the current predicted by Eq. (3.7) reaches saturation and reverse saturation (i.e., with the role of the drain played by the source) at $V_{DB} = V_P$ and $V_{SB} = V_P$, respectively, with V_P as given by Eq. (3.9).

(*c*) Extend the model into the saturation region; i.e., prove Eq. (3.10).

(*d*) Compare Eq. (3.10) to the strong inversion saturation model of Table 3.1.

3.21 Consider the general model in Eq. (3.13). Show that this model reduces to Eq. (3.11) in weak inversion, to Eq. (3.7) in strong inversion nonsaturation, and to Eq. (3.10) in strong inversion saturation (assume $m \simeq m_0$). To what equation does the model reduce in strong inversion reverse saturation (i.e., saturation with the roles of drain and source interchanged)?

3.22 Consider the general model in Table 3.4. Show that in weak inversion and in strong inversion it is consistent with the results in Table 3.1, assuming that, in the value for V_M in Table 3.2, $c = \ln [2n/(n-1)]$ and $\alpha \simeq n$.

3.23 A common expression used to model the effect of V_{GS} on the effective mobility is given in Sec. 3.6. Plot $\sqrt{I_D}$ versus V_{GS} for $\theta = 0$ and $\theta = 0.125$ V^{-1} for a device with $W/L = 1$, other parameters as given in Example 3.8, and $V_{SB} = 0$.

3.24 For the device of Example 3.8 with $W = 2$ μm, $L = 2$ μm, $V_{SB} = 0$, $V_{GS} = 3$ V and operation in saturation, calculate I_D at 27 and 50°C. Use (3.19) and (3.20), with $c_1 = 1.8$ and $c_2 = 2$ mV/°C.

3.25 Using the results from Sec. 3.8, show that under strong velocity saturation conditions, the drain current in strong inversion saturation becomes approximately proportional to the first power of $V_{GS} - V_T$ and approximately independent of L.

3.26 This problem is intended to give a feel for the influence of short-channel effects.

(*a*) Find the current of a transistor with $W = 40$ μm, $L = 10$ μm, and parameters as given in Examples 3.7 and 3.8, with $V_{GS} = 3$ V, $V_{SB} = 0$ V, and $V_{DS} = 3$ V.

(*b*) Repeat the calculations for $W = 4$ μm and $L = 1$ μm, including the effects of velocity saturation and threshold geometry dependence. Use the results in Sec. 3.8, with typical values given there for the constants involved.

3.27 Rewrite all equations in Tables 3.1 and 3.2 for the case of *p*MOS devices.

3.28 For the *p*MOS device of Example 3.11, plot I_D versus V_{DS} with V_{GS} as a parameter, for the following cases: (*a*) $V_{SB} = 0$ V and $V_{GS} = -2, -3, -4$, and -5 V; (*b*) $V_{SB} = -2$ V and $V_{GS} = -2$ and -3 V. To avoid entering the breakdown regions, $|V_{DB}| = |V_{DS} + V_{SB}|$ should not exceed 5 V.

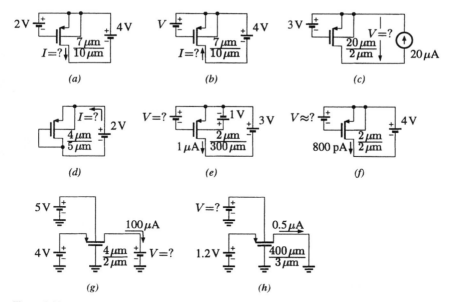

Figure 3.28

3.29   Calculate the quantities indicated in Fig. 3.28, assuming that $|V_A|$ is infinite and all other parameters are as given in Example 3.11. The quantities next to the devices, in the form a/b, mean $W = a$ and $L = b$.

3.30   The saturation drain current values of an nMOS transistor are measured at $V_{GS} = 2$ and 3 V. The results are as follows: 8.4 and 34 μA with $V_{SB} = $ 0 V; 5.30 and 28 μA with $V_{SB} = 1$ V; and 3.0 and 23 μA with $V_{SB} = 2$ V. Determine parameters K, V_{T0}, γ, and ϕ_0 for this device, using graphical techniques.

3.31   The method for graphically determining V_T described in Sec. 3.12 is valid if V_{DS} is negligibly small. Suggest a way to modify this method so that it can be used even in cases where V_{DS} cannot be neglected.

Chapter

4

MOSFET Small-Signal Modeling

4.1 Introduction

Having discussed dc large-signal operation in the previous two chapters, we will now assume that the terminal voltages of a transistor vary around their "bias" values by small amounts. The resulting current variations will then also be small, and they can be expressed in terms of the voltage variations by using linear relations. We will discuss such linear relations and will develop linear circuits to represent them [1]. Such circuits are called *small-signal equivalent circuits;* when excited by voltages equal to the *small variations* of the actual transistor voltages, these circuits will produce currents equal to the *small variations* of the actual transistor currents. Both conductance and capacitance parameters will be used as elements in these "models." We will assume that small-dimension effects are negligible, unless stated otherwise.

4.2 Small-Signal Conductance Parameters

Let us consider a MOS transistor biased with V_{GS}, V_{SB}, and V_{DS} fixed at values V_{GSQ}, V_{SBQ}, and V_{DSQ}, respectively, as shown in Fig. 4.1a. Let I_{DQ} be the resulting value of I_D. We can study the effect of very small changes of the terminal voltages on I_D by varying the voltages one at a time, as shown in Fig. 4.1b, c, and d. For now we are interested only in the change of the dc steady-state value of I_D; that is, we assume that the voltages are constant before and after each change and that I_D has reached dc steady state in both cases. We then consider the change ΔI_D between the two dc steady-state values. We can relate cause and effect by using three conductance parameters, which can be measured as shown next to each figure. These parameters are as follows:

1. *Small-signal gate transconductance* g_m, often referred to simply as *transconductance*. Mathematically, it is defined by the following relation, corresponding to the measurement in Fig. 4.1b:

$$g_m = \frac{\partial I_D}{\partial V_{GS}}\bigg|_Q \tag{4.1}$$

where the simplified notation used, with Q next to the vertical line, implies that the derivative must be evaluated at $V_{GS} = V_{GSQ}$, $V_{BS} = V_{BSQ}$, and $V_{DS} = V_{DSQ}$.

2. *Small-signal substrate transconductance* g_{mb}. Corresponding to the measurement in Fig. 4.1c, we have

$$g_{mb} = \frac{\partial I_D}{\partial V_{BS}}\bigg|_Q \tag{4.2}$$

Increasing V_{BS} as shown in Fig. 4.1c decreases V_{SB}. Consideration of the results of the body effect on I_D (Sec. 2.4.2) shows that I_D increases. Thus, ΔV_{BS} in Fig. 4.1c has on I_D qualitatively the same effect as ΔV_{GS} has in Fig. 4.1b. The substrate acts, in this sense, as a second gate, and it is often referred to as the *back gate*.

3. *Small-signal drain-source conductance* g_{ds}. Corresponding to the measurement in Fig. 4.1d, we define

$$g_{ds} = \frac{\partial I_D}{\partial V_{DS}}\bigg|_Q \tag{4.3}$$

The approximate equality signs in Fig. 4.1 become equality signs as ΔV_{GS}, ΔV_{BS}, and ΔV_{DS} approach zero. All three parameters have units of conductance and are often expressed in microsiemens (μS) or microamperes per volt (μA/V).

Let us now consider the general case in which all three voltages are changed simultaneously, as shown in Fig. 4.2. The corresponding total change in the drain current will be

$$\Delta I_D \simeq \frac{\partial I_D}{\partial V_{GS}}\bigg|_Q \Delta V_{GS} + \frac{\partial I_D}{\partial V_{BS}}\bigg|_Q \Delta V_{BS} + \frac{\partial I_D}{\partial V_{DS}}\bigg|_Q \Delta V_{DS} \tag{4.4}$$

From our above definitions, this becomes

$$\Delta I_D \simeq g_m \, \Delta V_{GS} + g_{mb} \, \Delta V_{BS} + g_{ds} \, \Delta V_{DS} \tag{4.5}$$

In addition, if, as before, the gate and substrate conductive currents are assumed to be fixed at zero (no leakage), we will have

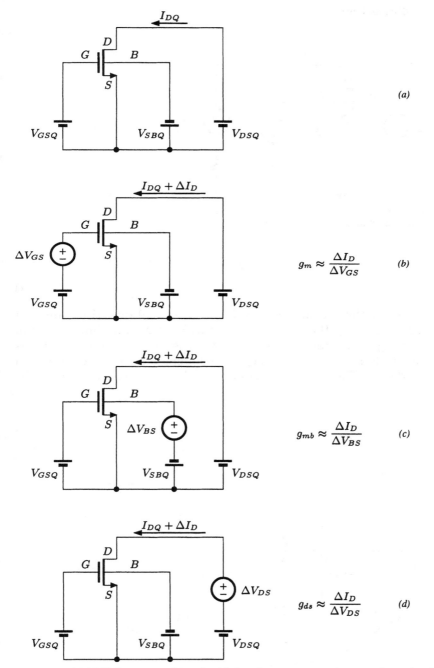

Figure 4.1 (a) A MOS transistor biased with dc voltages at a certain operating point. For this operating point, (b) shows the measurement of gate small-signal transconductance, (c) the measurement of body small-signal transconductance, and (d) the measurement of drain small-signal conductance. All voltage and current changes indicated by Δ are assumed to be very small.

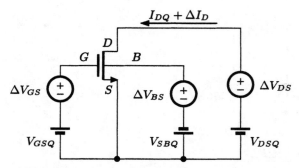

Figure 4.2 The MOS transistor of Fig. 4.1 with all three terminal voltage changes applied simultaneously.

$$\Delta I_G = 0 \qquad (4.6)$$

$$\Delta I_B = 0 \qquad (4.7)$$

The above three equations involve small-signal quantities and can be represented by the small-signal equivalent circuit shown in solid lines in Fig. 4.3, where the rhombic symbols represent controlled current sources. This circuit is widely used for small-signal calculations in circuit design. The extra conductance g_{db}, shown in broken lines, can be

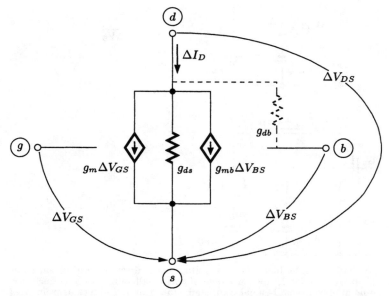

Figure 4.3 A low-frequency small-signal equivalent circuit for the MOS transistor.

added if the effect of impact ionization (Sec. 3.2.1, next to last paragraph) becomes considerable; the drain-substrate current I_{DB} caused by this effect varies with V_{DB}, and g_{db} is defined as the rate of change dI_{DB}/dV_{DB}.

In the preceding model derivation, the changes ΔV_{GS}, ΔV_{BS}, and ΔV_{DS} represent differences between two dc steady-state values of terminal voltages V_{GS}, V_{BS}, and V_{DS}, respectively. However, the model derived will be valid for representing the effects of the terminal voltages on the drain current, even if the changes are continuously varying with time, as long as the variations are slow enough that capacitive effects can be neglected. This will be understood better, and will be made more quantitative, after more complete models are considered.

For brevity, the words "small signal" may be omitted when we refer to a small-signal parameter when there is no chance for confusion; for example, g_{ds} will be referred to as the "drain conductance."

The above concepts and definitions are general—they were presented without regard to whether the transistor is nMOS or pMOS. Therefore they are valid for both device types, and so is the small-signal equivalent circuit of Fig. 4.3.

4.3 Expressions for Small-Signal Conductance Parameters in Weak and in Strong Inversion

For a given bias point, the quantities g_m, g_{mb}, and g_{ds} can be evaluated by using appropriate current expressions in the definitions of the previous subsection. For nMOS devices in weak or strong inversion, one can use the expressions in Tables 3.1 and 3.2 in this process. The general case, including moderate inversion, will be considered in Sec. 4.8. From now on, the subscript Q will be omitted for simplicity; it should be understood that the partial derivatives and the expressions developed from them are to be evaluated at the bias point of interest.

The results to be derived in this section will be approximate but simple, intended mainly for hand analysis. More accurate results can be derived at the expense of extra complexity; such results are used in computer simulation models [1–14].

4.3.1 Gate transconductance

In strong inversion saturation we have, from Table 3.1, $I_D = (k/2)(V_{GS} - V_T)^2$. Thus, from Eq. (4.1), the gate transconductance in this region will be

$$g_m = \frac{\partial}{\partial V_{GS}}\left[\frac{k}{2}(V_{GS} - V_T)^2\right] = k(V_{GS} - V_T) \qquad (4.8)$$

This can be put in alternative, useful forms by solving $I_D = (k/2)(V_{GS} - V_T)^2$ for $V_{GS} - V_T$ (or for k) and substituting in the above expression. This gives

$$g_m = k(V_{GS} - V_T) = \sqrt{2kI_D} = \frac{2I_D}{V_{GS} - V_T} \qquad (4.9)$$

The above results are summarized in Table 4.1, along with corresponding results for g_m in strong inversion nonsaturation and in weak inversion; these are easily derived in a similar manner. The expressions for other parameters in the table will be discussed shortly.

Let us now look at the behavior of the gate transconductance (a very important parameter) in some detail. As seen in Table 4.1, in strong inversion nonsaturation g_m is independent of V_{GS}. This is illustrated graphically in Fig. 4.4. We assume that the V_{GS} step used to obtain successive curves is fixed. The value of g_m can be estimated from the spacing of the curves (ΔI_D) if the V_{GS} steps (ΔV_{GS}) are small. As seen, for lines a and b the spacing is independent of V_{GS} but depends on V_{DS}. In

TABLE 4.1   Approximate Expressions for Small-Signal Conductances and Transconductances in Weak and in Strong Inversion (nMOS Transistor)

	Weak Inversion	
	Nonsaturation	Saturation
g_m		$\dfrac{I_D}{n\phi_t}$
g_{mb}		$(n-1)g_m$
g_{ds}	$\dfrac{I_{D,SAT}}{\phi_t} \exp\!\left(-\dfrac{V_{DS}}{\phi_t}\right)$ *	$\dfrac{I_D}{V_A}$ †
	Strong Inversion	
	Nonsaturation	Saturation
g_m	KV_{DS}	$\sqrt{2kI_D}$ ‡
g_{mb}	$\dfrac{\gamma}{2\sqrt{V_{SB} + \phi_0 + 0.4V_{DS}}}\, g_m$	$\dfrac{\gamma}{2\sqrt{V_{SB} + \phi_0 + 0.4V'_{DS}}}\, g_m$
g_{ds}	$K(V_{GS} - V_T - \alpha V_{DS})$ §	$\dfrac{I_D}{V_A}$ ¶

* $I_{D,SAT}$ is the current that would be obtained in *saturation* for the given V_{GS} and V_{SB}. The nonsaturation expression given for g_{ds} can become very inaccurate as saturation is approached. It should not be used if it predicts a value less than the saturation value of g_{ds}.
† This can be very inaccurate. See text.
‡ Alternate expressions for g_m in strong inversion saturation are $k(V_{GS} - V_T)$ and $2I_D/(V_{GS} - V_T)$.
§ Not valid if V_{DS} is within a few tenths of a volt of V'_{DS}. See Fig. 4.6b and associated discussion.
¶ This can be very inaccurate, especially outside a region $V''_{DS} < V_{DS} < V'''_{DS}$; see Fig. 4.6$b$ and associated discussion. Here V''_{DS} is above V_{DS} by 0.1 to 0.2 V.

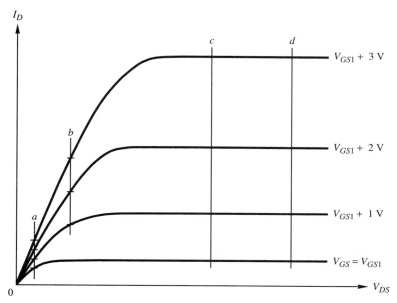

Figure 4.4   A family of I_D-V_{DS} curves, obtained for fixed V_{GS} increments. In non-saturation, the spacing is proportional to V_{DS} but independent of V_{GS} (lines a and b); in saturation, the spacing is independent of V_{DS} but depends on V_{GS} (lines c and d).

saturation, the situation is reversed, as illustrated by lines c and d. Now the spacing (and g_m) is independent of V_{DS} but depends on V_{GS}.

From Table 4.1 it follows that in strong inversion saturation, g_m, being equal to $\sqrt{2kI_D}$, is determined by both I_D and W/L (the latter enters through k; see Table 3.1). In contrast to this, in weak inversion g_m is equal to $I_D/(n\phi_t)$ and is fixed once I_D alone is fixed. This is a consequence of I_D being an exponential function of V_{GS} in weak inversion. In fact, the behavior here is qualitatively the same as for a bipolar junction transistor (Chap. 5), for which the transconductance is I_C/ϕ_t, with I_C being the collector current. The only difference is that in the case of the weakly inverted MOS transistor there is the additional factor $n > 1$ in the denominator, and thus the device exhibits a somewhat smaller transconductance than a bipolar transistor with the same current.

A plot of g_m versus I_D in saturation is shown in Fig. 4.5. Logarithmic axes are used, to accommodate the several orders of magnitude of change as one moves from weak, through moderate, to strong inversion (leakage currents are not considered). For each W/L value, the straight-line segment on the right corresponds to strong inversion; the middle, curved part corresponds to moderate inversion; and the straight-line segment on the left corresponds to weak inversion. As seen in strong

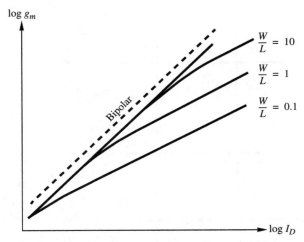

Figure 4.5 g_m versus I_D (both on log axes) for a MOS transistor with varius W/L values (solid lines) and for a bipolar transistor (broken line).

inversion, for a fixed value of I_D, one can obtain a large g_m by using a large value of W/L. However, if W/L is raised too much (always for a fixed current), weak inversion is entered and then further increases in W/L do not affect the value of g_m.

4.3.2 Body transconductance

The body transconductance g_{mb} can be calculated similarly from definition (4.2). Let us first consider a simplified case in strong inversion: Let us assume that α in the strong inversion expressions of Table 3.1 is independent of V_{SB} and is equal to 1. This is a common assumption which we have shown in Sec. 3.2 to be generally poor (unless γ is very small). Based on this assumption, we have $k = K$ and $I_D = (K/2)(V_{GS} - V_T)^2$ in saturation. This current depends on $V_{BS} = -V_{SB}$ indirectly; the dependence is hidden in V_T, which depends on V_{SB} due to the body effect. Thus the substrate transconductance will be

$$g_{mb} = \frac{\partial I_D}{\partial V_{BS}} = K(V_{GS} - V_T)\left(-\frac{dV_T}{dV_{BS}}\right)$$

$$= K(V_{GS} - V_T)\frac{dV_T}{dV_{SB}}, \qquad \alpha = 1 \qquad (4.10)$$

Using results from Table 3.2, we get

$$\frac{dV_T}{dV_{SB}} = \frac{\gamma}{2\sqrt{V_{SB} + \phi_0}} = n - 1 \tag{4.11}$$

Using this in Eq. (4.10) and observing that under our assumption of $\alpha = 1$ we have $K(V_{GS} - V_T) = g_m$, we obtain

$$g_{mb} = \frac{\gamma}{2\sqrt{V_{SB} + \phi_0}}\, g_m = (n - 1)g_m, \qquad \alpha = 1 \tag{4.12}$$

It is easily checked that the same expression is obtained in nonsaturation. Unfortunately, the above result is not accurate for most real devices, for which α is larger than unity and is a function of $V_{SB} = -V_{BS}$. This dependence must be taken into account when one is differentiating to obtain g_{mb}, and unfortunately it results in expressions too complicated for hand analysis of circuits. These expressions turn out not to be too accurate, despite their complexity. In general, reasonable accuracy in a function (in this case I_D) does not necessarily imply reasonable accuracy in a derivative of that function (here $\partial I_D/\partial V_{BS}$); we will see more illustrations of this important observation later. Owing to these problems, in Table 4.1 we suggest an alternate expression for g_{mb} which is simpler and rather accurate [1]; this expression is related to the accurate strong inversion I_D model given in Sec. 3.4 (Prob. 4.8). By comparison to this expression, it is obvious that Eq. (4.12) is mostly accurate for large V_{SB}, or small V_{DS}, or small $V_{GS} - V_T$ (and thus small V'_{DS}).

Similar considerations lead to the simple expression given in Table 4.1 for g_{mb} in weak inversion.

4.3.3   Drain-source conductance

The drain-source conductance g_{ds} can, in principle, be evaluated from its definition in Eq. (4.3). Caution is required here, especially in the saturation region. For example, the simple strong inversion saturation formula $I_D = (k/2)(V_{GS} - V_T)^2$ predicts $g_{ds} = \partial I_D/\partial V_{DS} = 0$, which is clearly incorrect; g_{ds} is simply the slope of I_D-V_{DS} curves like those in Fig. 2.11, and it is nonzero even in saturation. Thus we will use Eq. (3.3) instead; this gives

$$g_{ds} = \frac{\partial}{\partial V_{DS}}\left[\frac{k}{2}(V_{GS} - V_T)^2\left(1 + \frac{V_{DS}}{V_A}\right)\right]$$

$$= \frac{(k/2)(V_{GS} - V_T)^2}{V_A} \approx \frac{I_D}{V_A} \tag{4.13}$$

where the last approximation follows from assuming $V_{DS}/V_A \ll 1$.

When I_D is calculated, an error in the value of V_A is not disastrous, since usually $V_{DS}/V_A \ll 1$, and the error in the factor $1 + V_{DS}/V_A$ is small. However, since $g_{ds} = I_D/V_A$, an error in V_A will cause a comparable error in g_{ds}! This is an example of the important fact stated earlier: *A small error in a function does not necessarily imply a small error in a derivative of that function.*

A similar analysis is valid in weak inversion saturation. The result is included in Table 4.1, along with nonsaturation expressions easily derived from Table 3.1.

The nonsaturation expressions become inaccurate as saturation is approached, since they tend to predict too small a value for g_{ds}. This is now discussed with the help of Fig. 4.6. In Fig. 4.6a, the solid line is a plot for I_D versus V_{DS} in strong inversion for a real device; the upward trend at large V_{DS} values is due to short-channel effects (Sec. 3.8) and/or breakdown (Sec. 3.9). The broken line corresponds to the model of Table 3.1 (with infinite V_A). It is seen that the error in predicting I_D is small throughout. Consider now the slopes of the two plots, which give the corresponding g_{ds}; these are shown in Fig. 4.6b. A large error is now seen between the model (broken line) and the measurement (solid line), and it is yet another manifestation of the difference between the accuracy of a function and the accuracy of its derivative. The error becomes evident beginning at values of V_{DS} below V'_{DS} (usually by a few tenths of a volt). This problem is due to the fact that the drain current model is based on the assumption that the drain end of the channel is strongly inverted all the way up to $V_{DS} = V'_{DS}$, at which point it abruptly becomes pinched off (Sec. 2.4.3). Actually, though, before pinch-off is approached, the drain end of the channel becomes moderately inverted, a fact which is not taken into account in the model.

In saturation, the error in the modeled g_{ds} (broken line in Fig. 4.6b) is of course clearly unacceptable. This error is due, as we mentioned, to the zero slope of the broken line in Fig. 4.6a. Even if the expression $g_{ds} = I_D/V_A$ is used (Table 4.1), g_{ds} cannot be modeled accurately since that expression predicts a practically constant value (unless V_A is made a function of V_{DS}), as shown by the dotted-dashed line in Fig. 4.6b. At least, if the value of V_A is properly chosen, one can use I_D/V_A for rough estimates *provided* V_{DS} is in the range where the g_{ds} plot is relatively flat. In Fig. 4.6b, this range can be considered to be approximately between $V_{DS} = V''_{DS}$ and $V_{DS} = V'''_{DS}$. The value of V''_{DS} is usually 0.1 to 0.2 V above V'_{DS}. The value of V'''_{DS} depends on a variety of small-dimension and breakdown effects (Secs. 3.8 and 3.9) and is difficult to define and predict. For devices with short channels there may be no noticeable "flat" portion in the g_{ds} plot.

From the above discussion it is obvious that in calculating g_{ds} in saturation, or even in the transition from nonsaturation to saturation, a good computer simulator is indispensable. However, even in computer models the accuracy of g_{ds} is often poor. This will be discussed further in Sec. 4.9.

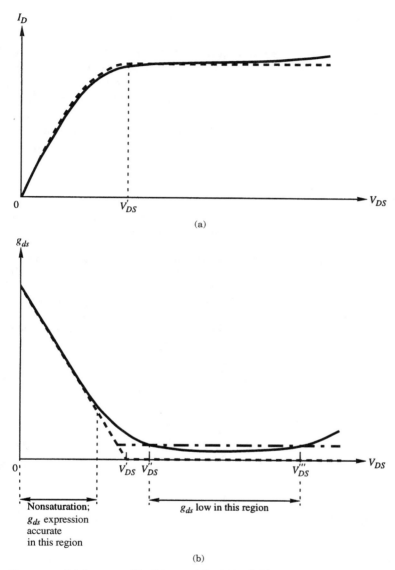

(a)

(b)

Figure 4.6 (a) I_D versus V_{DS} for a real device (solid line) and for our model, assuming infinite V_A (broken line); (b) slope g_{ds} corresponding to the curves in (a); the dashed-dotted line gives I_D/V_A, assuming V_A is constant.

4.3.4   Drain-substrate conductance

As already mentioned, a drain-substrate conductance g_{db} must sometimes be included in the small-signal model (Fig. 4.3) to take into account the effect of impact ionization. This conductance, which is difficult to model simply,* is included in some computer simulator models. Its effects may need to be considered in certain circuits which are intended to have very low output conductance, as the latter may be shunted by g_{db}. The value of g_{db} can be kept very low if V_{DS} is not allowed to exceed V'_{DS} by more than a couple of volts. In the rest of this chapter we will assume that g_{db} is negligible, unless we state otherwise.

4.3.5   Examples

We now try to get a feel for the magnitudes involved in the calculation of small-signal conductance parameters through an example.

Example 4.1   Calculation of small-signal conductances.   Calculate the small-signal conductance parameters for the nMOS device of Example 3.8 (for which $W = 10$ μm and $L = 3$ μm), biased with $V_{GS} = 3$ V, $V_{SB} = 2$ V, and $V_{DS} = 2.5$ V.

solution   The value of V_{SB} is the same as that used in Example 3.8b and c. Using quantities determined there in expressions from Table 3.1, we easily find that for the given V_{GS} we have $V'_{DS} = 1.77$ V, and the device operates in strong inversion saturation with $I_D = 437$ μA. From Table 4.1 we get

$$g_m = \sqrt{2(237 \ \mu\text{A/V}^2)(437 \ \mu\text{A})} = 455 \ \mu\text{A/V}$$

$$g_{mb} = \frac{0.35 \ \text{V}^{1/2}}{2\sqrt{2\text{V} + 0.76\text{V} + 0.4 \times 2.5\text{V}}} \times 455 \ \mu\text{A/V}$$

$$= 0.09 \times 455 \ \mu\text{A/V} = 41 \ \mu\text{A/V}$$

We have $V''_{DS} \approx V'_{DS} + 0.1$ V $= 1.87$ V (see Sec. 4.3.3). Since V_{DS} is larger than this value, we use the formula for g_{ds} from Table 4.1, to obtain

$$g_{ds} \approx \frac{437 \ \mu\text{A}}{63 \ \text{V}} \simeq 7 \ \mu\text{A/V}$$

We should be aware, though, that this is only a rough estimate, for reasons already explained.

Example 4.2   For the device of Example 4.1, assume that V_{GS} is decreased so that $I_D = 2$ nA. Find the small-signal conductance parameters.

solution   The maximum weak inversion current I_M was found in Example 3.8 (part c) to be 18 nA, so with $I_D = 2$ nA the device is in weak inversion saturation

---

* From the empirical model mentioned at the end of Sec. 3.2.1 [2], it can be estimated that $g_{db} \simeq I_{DB}V_I/(V_{DS} - 0.8V'_{DS})^2$, where I_{DB} is the drain-substrate impact ionization current and V_I is an empirical constant (typically 30 V). This formula is only adequate for order-of-magnitude calculations.

(note the large value of V_{DS} given). Using values from Example 3.8b in expressions from Table 4.1, we have

$$g_m = \frac{2 \text{ nA}}{1.105 \times 0.026 \text{ V}} \approx 70 \text{ nA/V}$$

$$g_{mb} = (1.105 - 1)(70 \text{ nA/V}) \approx 7.3 \text{ nA/V}$$

$$g_{ds} \approx \frac{2 \text{ nA}}{63 \text{ V}} = 0.032 \text{ nA/V}$$

where it should always be kept in mind that the result for g_{ds} is very approximate. In particular, if the value of V_A used was optimized for use in strong inversion, the weak inversion result may be especially poor.

The small-signal parameter expressions for pMOS devices can be derived in a similar manner. Such expressions can be obtained from those in Table 4.1 by using rather obvious sign changes. In engineering practice, one often follows the approach illustrated in Example 3.11. In this approach, the small-signal parameters of a pMOS device are equal to those found for the "corresponding" nMOS device.

Example 4.3 Find the small-signal conductance parameters for a pMOS device with $K = 257$ μA/V², $\gamma = 0.35$ V¹ᐟ², $V_{T0} = -0.8$ V, and $|V_A| = 63$ V, biased with $V_{GS} = -3$ V, $V_{SB} = -2$ V, and $V_{DS} = -2.5$ V.

solution This device corresponds exactly to the nMOS device of Example 4.1, and its bias corresponds to the bias used there. Thus, the results obtained there are valid here as well, and we have

$$g_m = 455 \text{ μA/V} \qquad g_{mb} = 41 \text{ μA/V} \qquad g_{ds} = 7 \text{ μA/V}$$

4.3.6 Small-signal parameters in the presence of second-order effects

If the equations in Table 3.1 are not valid due to the strong presence of other effects (e.g., effective mobility dependence on gate voltage, or short-channel effects), the formulas in Table 4.1 will not be valid either. One then has to develop different formulas by differentiating drain current formulas which include appropriate expressions, like the ones in Secs. 3.6 and 3.8. One case worth noting is that of velocity saturation. As indicated in Sec. 3.8, if this effect is strong, I_D becomes independent of L and varies linearly with V_{GS}, as indicated in Fig. 3.15b; thus g_m will also be independent of L and will be independent of V_{GS}, too.

4.4 Capacitance Parameters

A varying small-signal voltage $\Delta v(t)$ across two transistor terminals causes corresponding changes in the charges internal to the transistor. These must be supplied from the external world. Let $\Delta q(t)$ be the contribution to such charges coming through a given terminal. If the voltage variation is fast (occurring over a small time interval Δt), then dq/dt can be significant. Let us assume that the internal charges, and thus $\Delta q(t)$, follow "in step" with $\Delta v(t)$; this is called *quasi-static operation* and is a valid assumption unless the variations in $v(t)$ are extremely fast (see below). Under quasi-static operation, the small quantity Δq will be proportional to Δv; the constant of proportionality has dimensions of capacitance. Thus $dq/dt = (dq/dv)(dv/dt) = C(dv/dt)$ represents a *capacitive* current. Such capacitive currents will flow in addition to the currents predicted by the conductance parameters in Fig. 4.3, and their effect can be included by adding capacitors to the figure. This results in the model shown in Fig. 4.7, where v_{gs} and v_{bs} represent small-signal variations of the gate-source and substrate-source voltages, respectively.

To facilitate our discussion of capacitances, we will view the MOS transistor as consisting of two parts. The part between source and drain, containing the inversion layer and the depletion region, as well as the oxide and gate immediately above them, is called the *intrinsic* part; this is the part responsible for transistor action. This main part is surrounded by *parasitic* elements: the source and drain resistance, the junction capacitance between those structures and the body, the resistance of the body, etc. These undesirable (but also unavoidable) parasitic elements constitute the *extrinsic* part. In general, a total capacitance C_{xy} between terminals x and y can be expressed as the parallel combination of two capacitances:

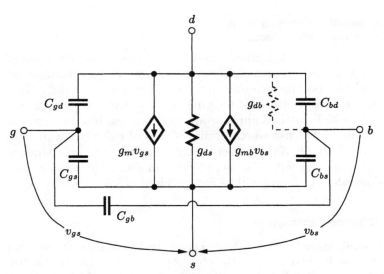

Figure 4.7 Small-signal equivalent circuit including capacitances.

$$C_{xy} = C_{xye} + C_{xyi} \qquad (4.14)$$

where C_{xye} represents the contribution of the extrinsic part and C_{xyi} the contribution of the intrinsic part. We will consider extrinsic and intrinsic contributions separately.

4.4.1 Extrinsic capacitances

To understand the origin of extrinsic capacitances, we need to consider the structures around the main, intrinsic transistor part in some detail. An nMOS transistor with its immediate surroundings is shown in Fig. 4.8, where a is the top view, b is a cross-sectional view along the channel's length, and c is a cross-sectional view along the channel's width. As seen, the channel boundary is defined by the inside walls of the source and drain (see Fig. 4.8b), as well by the boundary of the "thin" gate oxide (see Fig. 4.8c). As seen in Fig. 4.8c, on either side of the channel the oxide becomes thick (in practice it could be, for example, 40 times as thick as the thin oxide). The source-channel-drain combination is surrounded by a p^+ *channel-stop* region. This region helps prevent the formation of parasitic inversion layers outside the channel area. The gate completely overlaps the channel and extends outside it on all four sides; this overlap is unavoidable due to construction details and reliability considerations. More details on these and other features of device construction can be found in Chap. 5; here we only summarize the features pertinent to our discussion of extrinsic capacitances. The main extrinsic capacitances are noted in the figure, near the parts of the structure responsible for their presence. We now discuss each capacitance.

In Fig. 4.8b, C_{gse} and C_{gde} are "two-plate" capacitances due to the overlap between gate and source and between gate and drain, respectively. These capacitances are equal in symmetric devices. Their value can be roughly estimated by multiplying the oxide capacitance per unit area by the overlap area, the latter being proportional to W. Actually, their value turns out to be somewhat larger due to the electric field lines which fringe outside the immediate area of the overlap. Still, though, the result will be practically proportional to W; the constant of proportionality is a capacitance per unit of channel width, and it will be assumed to have been specified for a given fabrication process.

Another overlap capacitance, C_{gbe}, exists between gate and body (Fig. 4.8c). It is due to the gate-body overlap on either side of the channel and along its length, and it is practically proportional to the channel length. The constant of proportionality is a capacitance per unit of channel length; it can be calculated from knowledge about the thin-to-thick oxide tapering shown in the figure, and it will be assumed to have been specified.

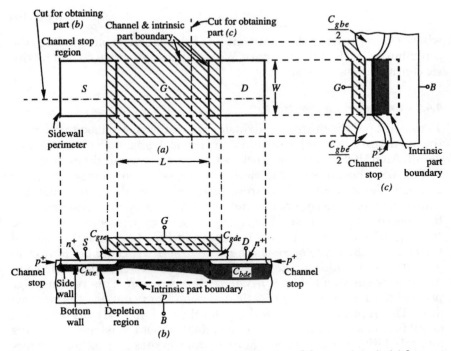

Figure 4.8 (a) Top view of a MOS transistor; (b) side view of the transistor in (a) for a cut along the length of its channel; (c) side view of the transistor in (a) for a cut along the width of its channel. Various extrinsic capacitances are indicated near the corresponding regions of the structure.

In addition to the overlap capacitances, there exist two *junction* capacitances: C_{bse} between body and source and C_{bde} between body and drain (Fig. 4.8b). Each of these consists of a *bottom-wall* part and a *sidewall* part. The reason we need to distinguish between these is that the presence of the heavily doped channel-stop region along the external perimeter of the source and drain makes the capacitance per unit area larger there. The sidewall capacitance is proportional to the "perimeter" along which it exists (Fig. 4.8a).* The constant of proportionality is a capacitance per unit of perimeter. This quantity can be calculated from knowledge of the shapes and doping profiles involved, and it will be assumed given (for zero junction bias). Also given, we assume, is the zero-bias bottom-wall capacitance per unit area. Like all

---

* It is rather common to include all four sides of the source or drain in the "perimeter." This is not strictly correct, as the internal side is not adjacent to the channel-stop area; however, proper consideration of the capacitance of that side is very involved. Including all four sides in the calculation tends to somewhat overestimate the capacitance.

junction capacitances, both the sidewall and bottom-wall parts of C_{bse} and C_{bde} depend on the junction reverse bias [1–4, 15].

Finally, as with any objects in proximity, a capacitance exists between the source and drain regions. Because of the complicated shapes involved, the value of this capacitance is difficult to evaluate. However, it is, in general, very small and can be neglected in comparison to other capacitances unless the channel is extremely short.

Expressions for the five main extrinsic capacitances are given in Table 4.2. These expressions can be easily seen to make sense based on what we have already discussed.

Example 4.4 Calculation of extrinsic capacitances. Assume that the device of Example 3.8 (for which $W = 10$ μm and $L = 3$ μm) has 10-μm by 3-μm rectangular source and drain regions. For the fabrication process used it is known that $C_w^* = 0.19$ fF/μm, $C_l^* = 0.3$ fF/μm, $C_{j0}' = 0.31$ fF/μm², $C_{j0}^* = 0.17$ fF/μm, $\phi_1 \approx \phi_2 = 0.8$ V, $\eta_1 = 0.5$, and $\eta_2 = 0.33$. Calculate the extrinsic capacitances for $V_{SB} = 2$ V, $V_{DS} = 2.5$ V, and $V_{GS} = 3$ V.

solution The source and drain areas are $A_S = A_D = 10$ μm × 3 μm = 30 μm², and their perimeters are $P_S = P_D = (10 + 3 + 10 + 3)$ μm = 26 μm. The junction biases are $V_{SB} = 2$ V and $V_{DB} = V_{DS} + V_{SB} = 4.5$ V. From Table 4.2 we have

TABLE 4.2 Expressions for Extrinsic Small-Signal Capacitances

Capacitance	Expression
C_{gse}	WC_w^*
C_{gde}	WC_w^*
C_{gbe}	LC_l^*
C_{bse}	$\dfrac{A_S C_{j0}'}{[1 + (V_{SB}/\phi_1)]^{\eta_1}} + \dfrac{P_S C_{j0}^*}{[1 + (V_{SB}/\phi_2)]^{\eta_2}}$
C_{bde}	$\dfrac{A_D C_{j0}'}{[1 + (V_{DB}/\phi_1)]^{\eta_1}} + \dfrac{P_D C_{j0}^*}{[1 + (V_{DB}/\phi_2)]^{\eta_2}}$

C_w^* = gate-source (or gate-drain) overlap capacitance per unit of channel width

C_l^* = gate-substrate overlap capacitance per unit of channel length (outside channel, counting both sides)

C_{j0}' = bottom-wall junction zero-bias capacitance per unit of bottom-wall *area*

ϕ_1 = bottom-wall junction built-in potential

η_1 = bottom-wall junction characteristic exponent

C_{j0}^* = sidewall junction zero-bias capacitance per unit of sidewall *perimeter*

ϕ_2 = sidewall junction built-in potential

η_2 = sidewall junction characteristic exponent

A_S = source bottom-wall junction area

A_D = drain bottom-wall junction area

P_S = source sidewall perimeter

P_D = drain sidewall perimeter

$$C_{gse} = C_{gde} = 10 \ \mu m \times 0.19 \ fF/\mu m = 1.9 \ fF$$

$$C_{gbe} = 3 \ \mu m \times 0.3 \ fF/\mu m = 0.9 \ fF$$

$$C_{bse} = \frac{30 \ \mu m^2 \times 0.31 \ fF/\mu m^2}{[1 + (2/0.8)]^{0.5}} + \frac{26 \ \mu m \times 0.17 \ fF/\mu m}{[1 + (2/0.8)]^{0.33}} = 7.9 \ fF$$

$$C_{bde} = \frac{30 \ \mu m^2 \times 0.31 \ fF/\mu m^2}{[1 + (4.5/0.8)]^{0.5}} + \frac{26 \ \mu m \times 0.17 \ fF/\mu m}{[1 + (4.5/0.8)]^{0.33}} = 6.0 \ fF$$

4.4.2 Intrinsic capacitances in weak and in strong inversion

Intrinsic capacitances are due to the charges inside the intrinsic part, which is shown enclosed by the broken line in Fig. 4.8b. These charges, and the resulting capacitance expressions, depend strongly on the region of operation. In weak inversion, the inversion layer charge is very small and does not play much of a role; the gate "looks right through it," directly at the body, as indicated schematically in Fig. 4.9a. Thus, the only significant intrinsic capacitance in weak inversion is that between the gate and the body C_{gbi} (we remind the reader that the subscript i stands for the *intrinsic* part of the total capacitance). It can be shown that this capacitance is equivalent to the series combination of the oxide capacitance and a capacitance corresponding to the depletion region below the oxide. The value of C_{gbi}, along with other results to be discussed shortly, is given in Table 4.3.

In strong inversion, the presence of the strong inversion layer gives rise to four additional intrinsic capacitances, as indicated schematically in Fig. 4.9b. A plot of the strong inversion intrinsic capacitances

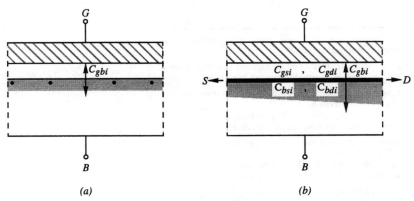

Figure 4.9 Part of a MOS transistor between source and drain that is pertinent to the discussion of intrinsic capacitance (a) in weak inversion and (b) in strong inversion.

TABLE 4.3   Approximate Expressions for Intrinsic Small-Signal Capacitances in Weak and in Strong Inversion

	Weak Inversion	
	$C_{gbi} = \dfrac{n-1}{n} WLC'_{ox},$ all others negligible*	

	Strong Inversion	
	Nonsaturation	Saturation
C_{gsi}	$\dfrac{WLC'_{ox}}{2}\left(1 + \dfrac{1}{3}\dfrac{V_{DS}}{V'_{DS}}\right)$	$\dfrac{2}{3}WLC'_{ox}$
C_{gdi}	$\dfrac{WLC'_{ox}}{2}\left[1 - \left(\dfrac{V_{DS}}{V'_{DS}}\right)^2\right]$	0
C_{gbi}	Negligible over most of region[†]	$\dfrac{\gamma}{2\sqrt{V_{SB}+\phi_0+V'_{DS}}}\dfrac{WLC'_{ox}}{3}$
C_{bsi}	$\dfrac{\gamma}{2\sqrt{V_{SB}+\phi_0}}\dfrac{WLC'_{ox}}{2}$ [‡§]	$\dfrac{\gamma}{2\sqrt{V_{SB}+\phi_0}}\dfrac{WLC'_{ox}}{2}$ [‡§]
C_{bdi}	$\dfrac{\gamma}{2\sqrt{V_{SB}+\phi_0}}\dfrac{WLC'_{ox}}{2}\left(1 - \dfrac{V_{DS}}{V'_{DS}}\right)$ [‡]	0

* Unless the channel is extremely long and the device is operating near the top of the weak inversion region (Sec. 4.8).

[†] C_{gbi} is 0 at $V_{DS} = 0$, remains negligible over most of the nonsaturation region, and as saturation is approached rises to its saturation value, given in the rightmost column.

[‡] Note that this is independent of C'_{ox}, since γ is inversely proportional to C'_{ox} (Table 3.3); it just is convenient to give the expression in terms of the familiar quantities γ and C'_{ox}.

[§] C_{bsi} decreases somewhat from the value shown, as V_{DS} is increased from 0 to V'_{DS}; this effect is neglected for simplicity.

versus V_{DS} is shown in Fig. 4.10.* We will briefly discuss these plots, starting from the case of $V_{DS} = 0$. Here a strong inversion layer exists throughout the channel. The total capacitance between it and the gate is the oxide capacitance per unit area, times the channel area from the source end to the drain end; i.e., it is equal to WLC'_{ox}. This is the total intrinsic gate capacitance $C_{gsi} + C_{gdi}$. By symmetry, $C_{gsi} = C_{gdi} = WLC'_{ox}/2$. In a similar manner, with $V_{DS} = 0$ by symmetry the capacitances C_{bsi} and C_{bdi} are equal, and each is given by one-half the total capacitance between the body and the inversion layer. As for C_{gbi}, its value is zero at $V_{DS} = 0$ (always assuming strong inversion), since the strong inversion layer acts as a "shield," preventing the body from "seeing" the gate.

Let us now look at the other extreme, *saturation.* Here the channel is assumed to be pinched off, as explained in Sec. 2.4.3. If V_{DS} is varied

* The magnitudes shown for the three substrate capacitances relative to the gate capacitances are typical; they may be larger or smaller than shown, depending on device details and bias.

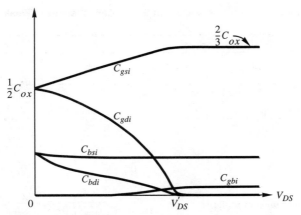

Figure 4.10 Intrinsic capacitances versus V_{DS} for fixed V_{SB} and V_{GS}. The relative size of the substrate capacitances in comparison to the gate capacitances will be different for different devices and/or biasing.

(always remaining above V'_{DS}), the structure to the left of the drain does not "feel" anything, since the pinched-off end remains at a potential of V'_{DS} with respect to the drain. Thus, V_{DS} has no effect on the intrinsic charges, just as it has no effect on the saturated drain current (neglecting channel length modulation and short-channel effects). Thus, $C_{gdi} = 0$ and $C_{bdi} = 0$.

The situation is different for the capacitances related to the source, though, since the source is strongly connected to the inversion layer. Thus C_{gsi} and C_{bsi} are nonzero even in saturation, as seen in Fig. 4.10. Nevertheless, one should not make the mistake of thinking of the inversion layer as a plate attached to the source because as we go toward the pinched-off channel end, the inversion level becomes lighter and lighter, and more and more electric field lines go from the gate, right through the inversion layer, to the depletion region below it. In other words, instead of the inversion layer's acting as a solid plate connected to the source (in which case the capacitance between it and the gate would be WLC'_{ox}), it acts as a plate with openings in it, especially as we get closer to the drain. This makes C_{gsi} less than the total oxide capacitance. A detailed derivation shows that $C_{gsi} = \frac{2}{3} WLC'_{ox}$.* A separate derivation shows that C_{bsi} can be assumed to be about equal to its value

---

* The factor $\frac{2}{3}$ is a result of a mathematical derivation [1] and has *nothing* to do with the lengths of the channel and of the pinch-off region; in fact, the pinch-off region length is assumed to be zero in such a derivation. Indeed, in devices with very long channels, the pinch-off region's length is a negligible part of L, but the factor of $\frac{2}{3}$ is still obtained in the expression for C_{gsi}.

at $V_{DS} = 0$.* As for C_{gbi}, its value is not zero in saturation due to the fact that some field lines go from the gate, through the channel, to the body; it is nevertheless small (typically a few percent of WLC'_{ox}).

At intermediate V_{DS} values, the capacitances can be estimated from the formulas in Table 4.3. Note that these results are approximate, representing the curves in Fig. 4.10 by simple equations. This is done in the interest of simplicity and speed in approximate hand calculations. More accurate results do exist [1], and they are included in computer simulators.

From the results presented so far, one can derive an interesting, approximate relation. At low V_{DS}, or low V_{GS}, or high V_{SB} values we have

$$\frac{C_{bsi}}{C_{gsi}} \approx \frac{C_{bdi}}{C_{gdi}} \approx \frac{g_{mb}}{g_m} \approx n - 1 = \frac{\gamma}{2\sqrt{V_{SB} + \phi_0}} = \frac{dV_T}{dV_{SB}} \qquad (4.15)$$

Combining the results of Tables 4.2 and 4.3 by using Eq. (4.14), one can calculate the total capacitances to be used in the small-signal model of Fig. 4.7. The case of moderate inversion is discussed in Sec. 4.8.

Example 4.5 Calculation of intrinsic capacitances. A device with $C'_{ox} = 1.73$ fF/μm^2 and the rest of its electrical parameters as in Example 3.8 has $W = 10$ μm and $L = 3$ μm and is biased with $V_{SB} = 2$ V, $V_{DS} = 2.5$ V, and $V_{GS} = 3$ V. Calculate its intrinsic capacitances.

solution It is easily checked that for the bias point given, the device is in strong inversion saturation and that $V'_{DS} = 1.77$ V. The total oxide capacitance is

$$WLC'_{ox} = 10 \ \mu\text{m} \times 3 \ \mu\text{m} \times 1.73 \ \text{fF}/\mu\text{m}^2 = 51.9 \ \text{fF}$$

Using this in the strong inversion saturation expressions from Table 4.3, we have

$$C_{gsi} = \frac{2}{3}(51.9 \ \text{fF}) = 34.6 \ \text{fF}$$

$$C_{gdi} = 0 \ \text{fF}$$

$$C_{gbi} = \frac{0.35 \ \text{V}^{1/2}}{2\sqrt{(2 + 0.76 + 1.77)} \ \text{V}} \times \frac{51.9 \ \text{fF}}{3} = 1.4 \ \text{fF}$$

$$C_{bsi} = \frac{0.35 \ \text{V}^{1/2}}{2\sqrt{(2 + 0.76)} \ \text{V}} \times \frac{51.9 \ \text{fF}}{2} = 2.7 \ \text{fF}$$

$$C_{bdi} = 0$$

Example 4.6 Complete small-signal model. Find the complete small-signal equivalent circuit (Fig. 4.7) for the device and bias point used in Example 4.1.

---

* It is actually somewhat smaller, due to the fact that the depletion region under the channel becomes deeper as we go toward the drain (compare Figs. 2.2e and 2.4e).

solution For this device we have found the extrinsic capacitances in Example 4.4 and the intrinsic capacitances in Example 4.5. Thus we have

$$C_{gs} = C_{gse} + C_{gsi} = 1.9\,\text{fF} + 34.6\,\text{fF} = 36.5\,\text{fF}$$

$$C_{gd} = C_{gde} + C_{gdi} = 1.9\,\text{fF} + 0\,\text{fF} = 1.9\,\text{fF}$$

$$C_{gb} = C_{gbe} + C_{gbi} = 0.9\,\text{fF} + 1.4\,\text{fF} = 2.3\,\text{fF}$$

$$C_{bs} = C_{bse} + C_{bsi} = 7.9\,\text{fF} + 2.7\,\text{fF} = 10.6\,\text{fF}$$

$$C_{bd} = C_{bde} + C_{bdi} = 6.0\,\text{fF} + 0\,\text{fF} = 6.0\,\text{fF}$$

The small-signal conductance parameters were calculated in Example 4.1. Using these and the above capacitances in the circuit of Fig. 4.7, we obtain the complete small-signal equivalent circuit of the given device at the given bias point, shown in Fig. 4.11.

4.5 Intrinsic Cutoff Frequency and Limits of Model Validity

Although at low frequencies the gate current is zero, at high frequencies a significant current can flow through the capacitances associated with the gate. Consider a transistor in saturation connected as in Fig. 4.12a. The voltage $\epsilon \sin \omega t$ is a sinusoidal small signal. The radian frequency ω (in radians per second) is given by $2\pi f$, where f is the frequency in hertz. The small-signal equivalent circuit for this connection can be derived by replacing the transistor by the model of Fig. 4.7 and substituting short circuits for all dc voltage sources (since for these sources

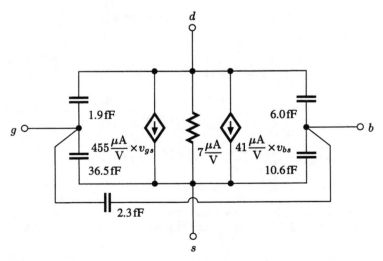

Figure 4.11 Small-signal equivalent circuit resulting from the calculation in Examples 4.1 and 4.6.

$\Delta v = 0$) [16]. Removing now all elements that appear in parallel with short circuits, neglecting extrinsic elements, and noting that C_{gdi} is zero in saturation, we arrive at the circuit of Fig. 4.12b. The small-signal drain and gate currents can be calculated by using this circuit, and they are shown directly on the figure. The amplitude of the drain current is $g_m\epsilon$; that of the gate current is $2\pi f(C_{gsi} + C_{gbi})\epsilon$. The *intrinsic cutoff frequency*, or *intrinsic unity current-gain frequency*, denoted by f_{Ti}, is defined as the value of f at which the amplitudes of the drain and gate currents become equal, that is, $g_m\epsilon = 2\pi f_{Ti}(C_{gsi} + C_{gbi})\epsilon$. Thus

$$f_{Ti} = \frac{g_m}{2\pi(C_{gsi} + C_{gbi})} \qquad (4.16)$$

In strong inversion, using g_m and C_{gsi} from Tables 4.1 and 4.3 and neglecting C_{gbi}, we obtain

$$f_{Ti} \simeq \frac{g_m}{2\pi C_{gsi}} \approx \frac{(\mu C'_{ox}/\alpha)(W/L)(V_{GS} - V_T)}{2\pi(2/3)WLC'_{ox}}$$

which gives

$$f_{Ti} \approx \frac{3}{2\alpha} \frac{\mu(V_{GS} - V_T)}{2\pi L^2}, \qquad \text{strong inversion saturation} \qquad (4.17)$$

This is only a rough estimate, because at frequencies as high as f_{Ti} the operation of the transistor is not truly quasistatic and our model is thus not really valid (see below). Since we used our model to predict the value of f_{Ti}, and since the model is not valid at $f = f_{Ti}$ in the first place, the predicted value of f_{Ti} should be viewed with suspicion. A more accurate value can be found from models valid up to higher frequencies (Sec. 4.6). The results turn out to be approximately the same as in Eq. (4.17); so, at least in this case, the quasistatic model did not let us down.

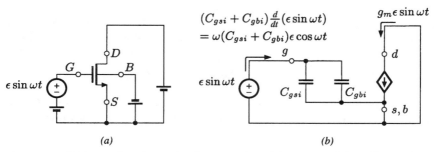

(a)   *(b)*

Figure 4.12 (*a*) A transistor operating in saturation with a sinusoidal small-signal excitation; (*b*) the small-signal equivalent circuit for (*a*).

Why calculate the intrinsic cutoff frequency? One reason is that it is a rough indication of the frequencies at which the intrinsic part of the device "gives up" in common amplification applications; in fact, in most integrated amplifiers the usable frequency range is well below the f_{Ti} of the devices used (this is not a hard rule, though; exceptions do exist). Note, after all, that the cutoff frequency of the *complete* device will be below f_{Ti} because of the effect of extrinsic parasitic capacitances and resistances, which we have not included in our calculation. The intrinsic cutoff frequency is then a figure of merit describing ideal behavior, which one strives to approach by minimizing all extrinsic parasitic elements as much as possible. As seen from Eq. (4.17), operation at high frequencies makes necessary the use of large gate-source bias and a small channel length.* Also, the presence of mobility in the above result indicates that for high-frequency applications nMOS devices are preferable, since their mobility is about 3 times higher than that of pMOS devices.

Another reason for calculating the intrinsic cutoff frequency is that it can be used in estimating the limits of validity of our small-signal model in Fig. 4.7. As can be found by comparison to nonquasistatic models (see Sec. 4.6), the model of Fig. 4.7 is only valid up to a fraction of f_{Ti}, for example, $0.5 f_{Ti}$ for rough calculations or $0.1 f_{Ti}$ for precision work (especially work involving precision calculations of phase). Thus a knowledge of the value of f_{Ti} allows us to estimate up to what frequency we can trust the results of circuit calculations using this model for a particular bias point of interest. The reader is warned that certain popular simulator models use this model at any frequency of operation, without providing a warning to the user when its limit of validity is exceeded.

Example 4.7   For the device we have been considering in the last several examples, $\mu = 445 \text{ cm}^2/(\text{V} \cdot \text{s})$ [$44.5 \text{ }\mu\text{m}^2/(\text{V} \cdot \text{ns})$], $V_{GS} = 3$ V, $V_T = 1.08$ V, and $L = 3$ μm. Thus

$$f_{Ti} \approx \frac{3}{2 \times 1.09} \frac{[44.5 \text{ }\mu\text{m}^2/(\text{V} \cdot \text{ns})](3 \text{ V} - 1.08 \text{ V})}{2\pi(3 \text{ }\mu\text{m})^2} = 2.1 \text{ GHz}$$

Thus the small-signal equivalent circuit shown in Fig. 4.11, developed for the device *and for the specific bias point considered*, will be valid up to a fraction of this frequency, for example, 1 GHz for simple, approximate calculations. However, if careful phase calculations are desired, the model should not be trusted above 200 MHz. Note that this is not likely to be a problem for many applications, in which the device would not be used at such high frequencies.

Let us now consider the value of f_{Ti} in weak inversion. Here C_{gsi} is negligible, but C_{gbi} is not. Using our weak inversion expressions for g_m

---

* Note, though, that if the channel length is made too small, velocity saturation will set in and the result in (4.17) will not be valid.

and C_{gbi} in Tables 4.1 and 4.3, respectively, we can show that (Prob. 4.15)

$$f_{Ti} \simeq \frac{g_m}{2\pi C_{gbi}} \approx \frac{\mu \phi_t}{2\pi L^2} \frac{I_D}{I_M}, \qquad \text{weak inversion} \qquad (4.18)$$

where I_D is the current at the particular operating point and I_M is the maximum possible weak inversion current (i.e., the saturation current at the upper limit of weak inversion; see Tables 3.1–3.3). The value of f_{Ti} here is much smaller than that in strong inversion; for example, for the device of the previous example with $I_D = I_M$, the above equation predicts $f_{Ti} = 20$ MHz, which is 2 orders of magnitude below the value found above for strong inversion. The reason is that although the transconductance is drastically decreased in weak inversion, the capacitance is not reduced all that much. Thus, pushing the high-frequency performance of MOS circuits makes necessary biasing in strong inversion.

4.6   The Transistor at Very High Frequencies

4.6.1   Extrinsic parasitic resistances

The source, drain, substrate, and gate regions are made of resistive materials and present a series resistance at each terminal (Chap. 5). Although in certain fabrication processes special techniques are used to lower series resistances, here let us consider a general case where the value of these resistances is not negligible. If the signals applied to a device are varying very fast, capacitance currents, which are of the form $C \, dv/dt$, can be large and can cause significant drops across these resistances.

Thus, at very high frequencies the effects of the resistances may not be negligible, in which case they should be included in the small-signal model. This is easier said than done, though, since these resistances are not lumped; for example, the resistance of the source region is *distributed* along with the junction capacitance of that region. Without appropriate simulation tools, one may have to use simple lumped approximations for these distributed elements. The simplest such approximation consists of using a resistor in series with each of the four terminals. More accurate (and more complicated) approaches are sometimes used [1–3].

4.6.2   Intrinsic nonquasi-static effects

The model of the intrinsic part of the transistor becomes inadequate at very high frequencies. It was said in the beginning of Sec. 4.4 that a key assumption in deriving the model of Fig. 4.7 is that the transistor operates *quasi-statically,* meaning that the charges in it change in step with the terminal voltages, displaying no "inertia." At very high speeds this is no longer true, and the device is said to operate *nonquasi-statically.* If, for example, the gate-source small-signal voltage is varying too fast, the inversion layer will exhibit inertia and will have difficulty responding. This has two effects. First, the resulting drain current change is no longer in phase with the gate-source voltage, but instead lags behind it. Second, the inertia of the inversion layer offers some resistance to what would otherwise be the capacitive gate current; to put it another way, the inversion layer resistance appears in series with the gate capacitance, and the resulting small-signal voltage drop across it cannot be neglected because the small-signal gate current is now significant.

The above effects can be included in the small-signal model. For example, a first-order model for the *intrinsic* part of a MOS transistor in strong inversion saturation is shown in Fig. 4.13a [17, 1]. For simplicity we assume that the extrinsic parasitic elements are negligible and that the source is connected to the substrate. Here we use current and voltage phasors [16, 18] corresponding to current and voltage waveforms in the sinusoidal steady state. As seen, in place of the transconductance we have a *transadmittance,* y_m. The denominator in y_m causes I_d to lag behind V_{gs}. This is a manifestation of the inversion layer's inertia, as is the presence of the resistance shown in series with the gate capacitance, according to the above discussion; note that this resistance should *not* be confused with the extrinsic series gate resistance mentioned earlier. The value of the characteristic frequency f_x given in the figure is easily seen to be 2.5 times the intrinsic cutoff frequency f_{Ti}, found in Eq. (4.17). For the device of Example 4.7, f_x is about 5.2 GHz at the bias point considered there.

The magnitude and phase of y_m are plotted versus frequency in Fig. 4.14 (solid curves). Clearly, such behavior is not predictable by the common quasi-static model of Fig. 4.7, for which $y_m = g_m$ (a quantity with constant magnitude and zero phase) by definition.

The model in Fig. 4.13a can be modified to avoid the use of a complex y_m, as shown in Fig. 4.13b. Here y_m has been replaced by g_m, but the current source depends not on V_{gs}, but rather on V_1; the latter is obtained from V_{gs} through a "low-pass" network consisting of R_1 and C_1, with R_1C_1 chosen so that the network has a cutoff frequency of f_x, making g_mV_1 equal to y_mV_{gs} in Fig. 4.13a. The individual values of R_1 and C_1 are chosen so that their series combination has a much larger impedance than the branch in parallel with it, and thus it does not modify the input impedance appreciably. We remind the reader that extrinsic elements are not shown in Fig. 4.13, but they should be included in a complete model.

Use of a first-order nonquasi-static model, like the ones in Fig. 4.13, extends the upper frequency limit of model validity by about 1 order of

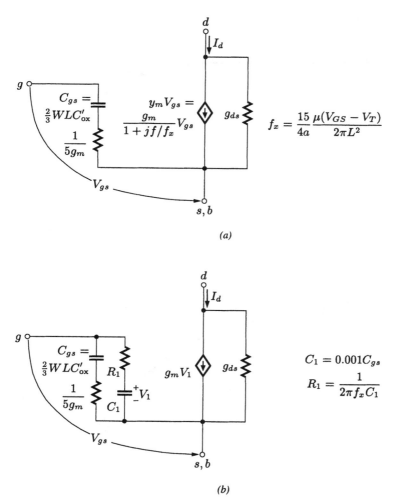

Figure 4.13 (*a*) Equivalent circuit for the intrinsic part of a MOS transistor in strong inversion saturation, including first-order nonquasi-static effects; (*b*) an equivalent model which does not need the specification of a complex y_m.

magnitude (to about f_{Ti}, for accurate work). At very high frequencies, though, even this model fails; a comparison to more accurate models [1] or measurements (broken line) is indicated in Fig. 4.14.

It is important to point out that the topology and values in the model of Fig. 4.13 have not been generated empirically; they result from careful derivations, in which the channel is viewed as a lossy transmission line. Gate-drain effects in nonsaturation as well as substrate effects can be taken into account in a similar way. Detailed discussions of non-quasi-static models can be found in the literature [1, 19–21]. Unfortu-

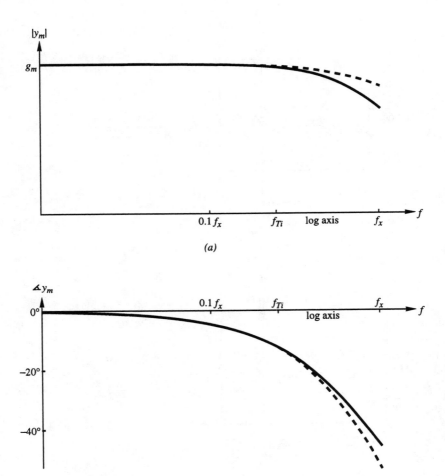

Figure 4.14 (*a*) Magnitude and (*b*) phase of MOSFET transadmittance. Solid lines are the model of Fig. 4.13, and broken lines are a high-order model.

nately, at the time of this writing, most computer simulator models have yet to incorporate even the simplest nonquasi-static models.

4.7 Noise

4.7.1 Introductory remarks

It has so far been assumed that the drain current of a MOS transistor varies with time only if one or more of the terminal voltages vary with time. This is not exactly true. A careful examination of the current reveals minute fluctuations, referred to as *noise,* which are there regardless of whether externally applied signals are present. Such fluctuations can occur owing to several mechanisms (see below). Noise can interfere with weak signals when the transistor is part of an analog circuit, so it is important to have available ways to predict it.

Consider a transistor with dc bias voltages, as shown in Fig. 4.15*a.* The total drain current, shown in Fig. 4.15*b,* can be expressed as

$$i_D(t) = I_D + i_n(t) \tag{4.19}$$

where I_D is the ideal (bias) current and $i_n(t)$ is the noise component, which has zero average value. The instantaneous value of i_n at a given t is, of course, unpredictable. Instead, one talks about certain measures characterizing the behavior of $i_n(t)$. In noise work, such measures are the *mean square* value, denoted by $\overline{i_n^2}$, and the *root mean square (rms)* value $\sqrt{\overline{i_n^2}}$, often denoted by $i_{n,\mathrm{rms}}$. For a noise voltage v_n, one can similarly define a mean square value $\overline{v_n^2}$ and an rms value $v_{n,\mathrm{rms}} = \sqrt{\overline{v_n^2}}$.

In calculating the total mean square noise due to several *independent* noise sources, one can consider the effect of each separately and then *add the individual mean square values.* To see why, consider a noise current $i_n(t)$ which consists of two independent contributions $i_{n1}(t)$ and $i_{n2}(t)$. We have $i_n(t) = i_{n1}(t) + i_{n2}(t)$, which gives $i_n^2(t) = i_{n1}^2(t) + i_{n2}^2(t) + 2i_{n1}(t)i_{n2}(t)$. The average of $i_n^2(t)$ is $\overline{i_n^2(t)} = \overline{i_{n1}^2(t)} + \overline{i_{n2}^2(t)} + \overline{2i_{n1}(t)i_{n2}(t)}$, where bars denote averages. However, since the two contributions are independent, their product averages to zero and thus

$$\overline{i_n^2} = \overline{i_{n1}^2} + \overline{i_{n2}^2} \tag{4.20}$$

or, in rms quantities

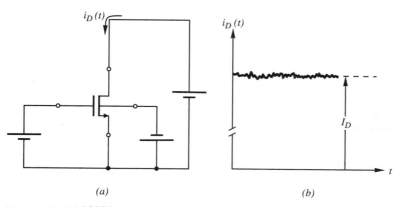

(a) (b)

Figure 4.15 (a) A MOS transistor biased with fixed noiseless terminal voltages; (b) the drain current for the connection in (a), including transistor-generated noise.

$$i_{n,\mathrm{rms}} = \sqrt{i_{n1,\mathrm{rms}}^2 + i_{n2,\mathrm{rms}}^2} \qquad (4.21)$$

In measuring noise quantities, the amount of noise depends on the bandwidth of the measuring instrument. A common measurement involves a *very narrow* bandwidth Δf, centered on a frequency f. The current noise spectral components within this bandwidth have a certain mean square value. The *ratio* of this value to Δf, as Δf is allowed to approach zero, tends to what is called the *power spectral density** of the current noise, denoted by $S_I(f)$. This quantity has units of square amperes per hertz. Often the square root of the power spectral density is used instead, given in $\mathrm{A}/\sqrt{\mathrm{Hz}}$. For a noise voltage v_n, one can similarly define a power spectral density $S_V(f)$ (in square volts per hertz) or its square root (in $\mathrm{V}/\sqrt{\mathrm{Hz}}$).

Let us now consider a noise current which is the sum of two independent noise currents. Consider the spectral components of these within a very small bandwidth Δf, and let i_{n1}^2, i_{n2}^2, and $\overline{i_n^2}$ be the corresponding mean square values of the noise components and their sum within that bandwidth. Equation (4.20) is valid; dividing by Δf and allowing Δf to approach zero, we easily find for the power spectral density of the sum

$$S_I(f) = S_{I1}(f) + S_{I2}(f) \qquad (4.22)$$

where $S_{I1}(f)$ and $S_{I2}(f)$ are the power spectral densities of i_{n1} and i_{n2}, respectively.

The total mean square noise current within an arbitrary bandwidth extending from $f = f_1$ to $f = f_2$ can be found by summing the mean square values of the individual components within each subbandwidth Δf. More precisely, using the power spectral density concept, we have

$$\overline{i_n^2} = \int_{f_1}^{f_2} S_I(f)\, df \qquad (4.23)$$

Similar results can be obtained for voltage noise.

A well-known example of device noise is the *thermal noise* in a resistor (also called *Johnson noise* or *Nyquist noise*). It is due to the random thermal motion of the carriers in it. A real (noisy) resistor (Fig. 4.16a) can be represented as an ideal (noiseless) resistor R in series with a noise voltage source as shown in Fig. 4.16b. The corresponding power spectral density is [23, 24]

$$S_{Vt}(f) = 4\hat{k}TR \qquad (4.24)$$

---

* More rigorous definitions can be found in texts dealing with stochastic processes [22].

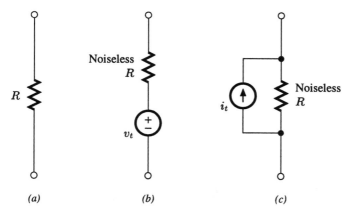

Figure 4.16 (a) A real (noisy) resistor; (b) Thevenin noise equivalent circuit for (a); (c) Norton noise equivalent circuit for (a).

where $\hat{k}$ is Boltzmann's constant (1.38×10^{-23} V · C/K) and T is the absolute temperature.* The above formula is valid up to extremely high frequencies (over 10^{12} Hz).

The circuit in Fig. 4.16b is a Thevenin equivalent circuit. This can be converted to a Norton equivalent circuit, shown in Fig. 4.16c, with rms current $i_{t,\mathrm{rms}} = v_{t,\mathrm{rms}}/R$, or equivalently with $\overline{i_t^2} = \overline{v_t^2}/R^2$; thus, the power spectral density of this noise current is

$$S_{It}(f) = 4\hat{k}T\,\frac{1}{R} \qquad (4.26)$$

4.7.2 MOS transistor noise

A typical plot of power spectral density for the drain current noise of a MOS device is shown in Fig. 4.17 on log-log axes. Two distinct frequency regions, with different noise behavior in each, can be identified. These regions can be thought of as separated by a *corner frequency f_c*. Values from several hertz to several hundred kilohertz are common for this quantity depending on device construction, geometry, and bias.

The type of noise dominating at high frequencies in Fig. 4.17 is *white noise* (its power spectral density is constant up to extremely high frequencies). In strong inversion, this component is actually *thermal noise* and is due to the random thermal motion of the carriers in the inversion layer. In weak inversion, the white noise component has been variously

---

* A convenient formula for $4\hat{k}T$ is

$$4\hat{k}T = 1.66 \times 10^{-20}\,\frac{\mathrm{V \cdot A}}{\mathrm{Hz}}\;\times\;\frac{\mathrm{T}}{300\ \mathrm{K}} \qquad (4.25)$$

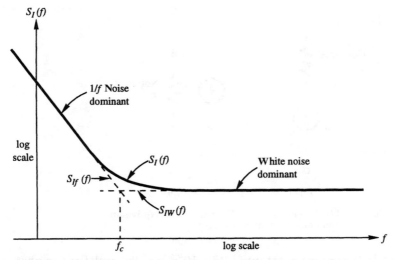

Figure 4.17 A typical plot of the power spectral density for the drain current noise versus frequency in log-log axes.

attributed to thermal noise or shot noise (a form of noise associated with dc current flow and with the ensuing discreteness of the charges carrying the current). Both theories produce identical results [25, 1, 2].

The noise dominating at low frequencies in Fig. 4.17 is called $1/f$ noise because the power spectral density for the current caused by this type of noise is practically proportional to $1/f$; it is also called *flicker noise*. This type of noise is often attributed to the presence of "traps" at the oxide-silicon interface, which can randomly exchange carriers with the channel and cause random fluctuations in its charge density. Another theory attributes $1/f$ noise to mobility fluctuations.

The white and $1/f$ noise components are independent. Let us denote their corresponding power spectral densities by $S_{Iw}(f)$ and $S_{If}(f)$, respectively. Following Eq. (4.22), we have

$$S_I(f) = S_{Iw}(f) + S_{If}(f) \qquad (4.27)$$

Expressions for the white and $1/f$ components in the right-hand side of this equation are given in Table 4.4 [1, 2, 23–25]. At $V_{DS} = V'_{DS}$, the nonsaturation formulas in Table 4.4 can be easily shown to reduce to the saturation expressions* (Prob. 4.17). In the $1/f$ noise expression, h

---

* In circuit literature and simulators, the saturation formulas are sometimes used indiscriminately in both saturation and nonsaturation. This can lead to very wrong results. For example, at $V_{DS} = 0$ we have $g_m = 0$, which would predict zero noise if those formulas were used! This is, of course, clearly false and physically impossible, as follows from derivations of the nonsaturation formulas in the table [1].

is a process-dependent quantity inversely proportional to $C_{ox}'^{a}$, where a is usually taken equal to 1 or 2. In general, cleaner fabrication processing results in lower values of h. The value of h for some devices is practically bias-independent, whereas for others, it tends to increase somewhat with V_{GS}. For a clean fabrication process with an oxide thickness of 200 Å, a typical value of h is 2×10^{-9} V$^2 \cdot \mu$m^2 for nMOS devices. For pMOS devices h is usually much smaller, a typical value for the above process being 8×10^{-11} V$^2 \cdot \mu$m^2.

The formulas in Table 4.4 do not include short-channel effects. The presence of the latter (Sec. 3.8) will modify the magnitude of the noise to some extent. In fact, if high electric fields are created within short-channel devices, the resulting high-energy carriers (hot carriers) can make the noise increase considerably.

Based on Eq. (4.19), a simple equivalent circuit for a real (noisy) transistor (Fig. 4.18a) is shown in Fig. 4.18b.* It consists of an ideal, *noiseless* transistor and a noise current source i_n, the power spectral density of which obeys Eq. (4.27) and Table 4.4. This current source is meant to be directly in parallel with the channel. Thus, if it is desired to include extrinsic source and drain parasitic resistances in the model, they should be added external to this circuit; needless to say, such parasitic resistances will contribute some noise of their own.

---

* This does not model "induced gate noise," i.e., noise coupled from the noisy channel to the gate terminal at very high frequencies, through the gate capacitance [24].

TABLE 4.4   Power Spectral Densities for Drain Current Noise in Weak and in Strong Inversion (nMOS Transistor)

	White Noise: $S_{Iw}(f)$*	
	Nonsaturation	Saturation
Weak inversion	$2qI_{D,SAT}\left[1 + \exp\left(-\dfrac{V_{DS}}{\phi_t}\right)\right]$ †	$2qI_D = 4\hat{k}T\left(\dfrac{1}{2}ng_m\right)$
Strong inversion	$(4\hat{k}T)K(V_{GS} - V_T)\left(1 - \dfrac{1}{3}\dfrac{V_{DS}}{V_{DS}'}\right)$	$4\hat{k}T\left(\dfrac{2}{3}\alpha g_m\right)$
	$1/f$ Noise: $S_{If}(f)$	
All regions	$\dfrac{hg_m^2}{WL}\dfrac{1}{f}$ ‡	

* All symbols in the above formulas have been encountered in Tables 3.1 to 3.3. The quantity $\hat{k}$ is Boltzmann's constant (1.38×10^{-23} V$\cdot$C/K); a convenient formula for $4\hat{k}T$ is

$$4\hat{k}T = 1.66 \times 10^{-20} \, \frac{\text{V} \cdot \text{A}}{\text{Hz}} \times \frac{T}{300 \text{ K}}$$

† $I_{D,SAT}$ is the current that would be obtained in saturation for the value of V_{GS} used.
‡ Parameter h is a fabrication-dependent parameter (see text) which for some devices is practically bias-independent but for others tends to increase somewhat with V_{GS}.

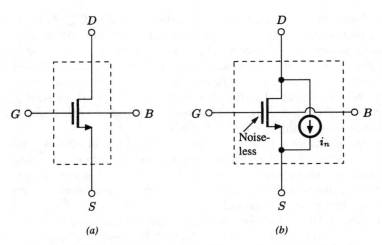

Figure 4.18   (*a*) A real (noisy) transistor; (*b*) an equivalent circuit for (*a*).

Example 4.8   Calculate the power spectral densities for the drain current white and $1/f$ noise as well as the corner frequency f_c (Fig. 4.17), for an nMOS device operating in strong inversion saturation with $V_{GS} - V_T = 2$ V, $W = 200$ μm, $L = 5$ μm, $k' = 68$ μA/V^2, $\alpha = 1.13$, and $h = 2 \times 10^{-9}$ V$^2 \cdot$μm^2. Calculate the contributions of each noise component, as well as the total noise, in the band from 10 Hz to 10 MHz.

solution   From Table 4.1 we have $g_m = k(V_{GS} - V_T) = 5440$ μA/V; thus, from Table 4.4 we have

$$S_{Iw}(f) = 1.66 \times 10^{-20}\,\frac{\text{V}\cdot\text{A}}{\text{Hz}} \times \frac{2}{3} \times 1.13 \times 5.44 \times 10^{-3}\,\frac{\text{A}}{\text{V}} = 6.8 \times 10^{-23}\,\frac{\text{A}^2}{\text{Hz}}$$

$$S_{If}(f) = \frac{2 \times 10^{-9}\,\text{V}^2 \cdot \text{μm}^2}{5 \times 200\,\text{μm}^2}\left(5.44 \times 10^{-3}\,\frac{\text{A}}{\text{V}}\right)^2\!\left(\frac{1}{f}\right) = 5.9 \times 10^{-17}\text{A}^2 \times \frac{1}{f}$$

The corner frequency f_c in Fig. 4.17 is the frequency at which the two contributions become equal; thus, equating the above two results and solving for $f = f_c$, we find

$$f_c = 868 \text{ kHz}$$

Let us denote the white and $1/f$ contributions to the noise current in the band from 10 Hz to 10 MHz by i_w and i_f, respectively. Integrating the corresponding power spectral densities, we find

$$\overline{i_w^2} = \int_{10\,\text{Hz}}^{10\,\text{MHz}} 6.8 \times 10^{-23}\,\frac{\text{A}^2}{\text{Hz}}\;df = 6.8 \times 10^{-16}\,\text{A}^2$$

$$\overline{i_f^2} = \int_{10\,\text{Hz}}^{10\,\text{MHz}} \frac{5.9 \times 10^{-17}\text{A}^2}{f}\;df = 8.2 \times 10^{-16}\,\text{A}^2$$

The total noise current in the given frequency band will have a mean square value of

$$\overline{i_n^2} = \overline{i_w^2} + \overline{i_f^2} = (6.8 \times 10^{-16} + 8.2 \times 10^{-16})\,\text{A}^2 = 1.5 \times 10^{-15}\,\text{A}^2$$

Thus the rms value of the noise current is

$$i_{n,\mathrm{rms}} = \sqrt{1.5 \times 10^{-15}\mathrm{A}^2} = 39 \text{ nA}$$

A common practice is to omit the symbol *rms* from the subscript and add it instead next to the units. Thus, the above result can be written as follows:

$$i_n = 39 \text{ nA rms}$$

Equivalent input noise voltage. Another common representation of transistor noise involves the *equivalent input noise voltage*. This quantity is defined as the noise needed in the voltage between the gate and source of a hypothetical *noiseless* transistor, to produce the correct amount of noise current (Fig. 4.19). Let us denote by v_n the equivalent input noise voltage. Assuming identical bias voltage sources are connected between gate and source in each of Figs. 4.18b and 4.19, the drain noise current must be the same in both for the two representations to be equivalent. Recalling the definition of the gate transconductance g_m, we thus must have $i_n = g_m v_n$ and $\overline{i_n^2} = g_m^2 \overline{v_n^2}$. Considering the components within a narrow bandwidth Δf, dividing by Δf, and allowing Δf to approach zero, we obtain

$$S_V(f) = \frac{1}{g_m^2} S_I(f) \tag{4.28}$$

The concept of the equivalent input noise voltage is fine at low frequencies, but it can lead to problems in high frequencies. Assume that the gate of the transistor in Fig. 4.19 is driven by a source with a high output impedance. Then v_n will be dropped partly across this impedance and partly across the transistor input capacitance. The fictitious, noiseless transistor will then see a noise voltage less than v_n, and the predicted noise in the current will be artificially low. Also, independent of frequency, the concept is not convenient at $V_{DS} = 0$, since then $g_m = 0$ and S_V in Eq. (4.28) can become infinite.

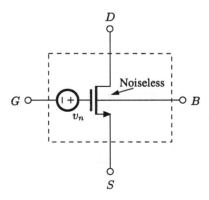

Figure 4.19 An equivalent circuit for a real (noisy) transistor, using the idea of the equivalent input noise voltage. This idea is only valid under certain conditions—see text.

The power spectral densities for the equivalent input noise voltage can be derived as suggested by Eq. (4.28) from the expressions of Table 4.4. Such results are given in Table 4.5.

Example 4.9 For the device of Example 4.8, calculate the power spectral densities of the white and $1/f$ equivalent noise voltage, and the square root of the power spectral density at 1 kHz (a commonly quoted quantity).

solution Using Table 4.5, or dividing the results of the previous example by g_m^2 [see Eq. (4.28)], we find

$$S_{Vw}(f) = 2.3 \times 10^{-18}\ \mathrm{V^2/Hz}$$

$$S_{Vf}(f) = (2 \times 10^{-12}\ \mathrm{V^2})\left(\frac{1}{f}\right)$$

To determine the square root of the power spectral density at 1 kHz, note that 1 kHz is much lower than the value of the corner frequency $f_c = 868$ kHz found in the previous example; thus, it can be assumed that the $1/f$ noise is completely dominant at 1 kHz. Using the result obtained above, we find $S_{Vf}(1\ \mathrm{kHz}) = 2 \times 10^{-15}\ \mathrm{V^2/Hz}$; the square root of this is

$$\sqrt{S_{Vf}(1\ \mathrm{kHz})} = 45\ \mathrm{nV}/\sqrt{\mathrm{Hz}}$$

4.8 General Models and Moderate Inversion

We have already seen in Sec. 3.5 that simple weak or strong inversion models for the drain current fail in the moderate inversion region. For this reason and for reasons of continuity, we found it necessary to describe there a general model for the drain current, valid in all regions

TABLE 4.5 Power Spectral Densities for Equivalent Input Voltage Noise in Weak and in Strong Inversion* (*n*MOS Transistor)

	White Noise: $S_{Vw}(f)$*	
	Nonsaturation	Saturation
Weak inversion	†	$4\hat{k}T\left(\dfrac{1}{2}n\dfrac{1}{g_m}\right)$
Strong inversion	†	$4\hat{k}T\left(\dfrac{2}{3}\alpha\dfrac{1}{g_m}\right)$

	$1/f$ Noise: $S_{Vf}(f)$
All regions	$\dfrac{h}{WL}\dfrac{1}{f}$

* The equivalent input noise voltage concept is valid only if the output impedance of the circuit driving the gate is much smaller than the gate input impedance. This may not be a valid assumption for certain circuits at high frequencies (see text). For the quantities involved in the formulas, see the footnotes in Table 4.4.

† The nonsaturation formulas for white noise can be produced by dividing the corresponding formulas in Table 4.4 by formulas for the nonsaturation g_m^2 (see text). The resulting expressions are rather complicated and are not given here since they predict infinite spectral density at $V_{DS} = 0$ (see text).

of inversion [25–28]. Expressions for the small-signal parameters corresponding to that model can be derived [27, 28]. Some of these are rather complicated. If we limit ourselves to the saturation region, though, simple expressions result that are valid in all regions of inversion. These are given in Table 4.6. Most of the expressions are obtained from Ref. 27. For completeness, an expression for the current, modified from one in Ref. 27, is also given [29]; this expression is discussed in Sec. 3.5.

A comparison of g_m/I_D as obtained from the general model of Table 4.6 and from weak and strong inversion expressions is shown in Fig. 4.20 [26, 27].* As seen, the latter expressions can be in rather large error.

The intrinsic capacitances change smoothly from their weak inversion values to their strong inversion values, as V_{GS} is raised. This is illustrated in Fig. 4.21. The variation seen in strong inversion is observed in real devices, but was not included in our tables for simplicity. The noise power spectral densities in moderate inversion also vary continuously from their weak inversion values to their strong inversion values, as V_{GS} is increased.

It was seen in Sec. 3.5 that most computer simulation models do an inadequate job in moderate inversion. This was illustrated in Fig. 3.10, which is reproduced here for convenience as Fig. 4.22a. The resulting inaccuracies in the small-signal parameters can be severe, as illustrated in Fig. 4.22b. The kink in the current makes the gate transconductance artificially jump as V_{GS} is raised past it, as shown. This is a

---

* In actual devices, as the current is lowered, g_m/I_D increases, peaks as weak inversion is entered, and then decreases slightly. This is because calculating g_m/I_D accentuates minute deviations from exponentiality. Such deviations can be seen to occur in the model of Eqs. (3.11) and (3.12), since m is a weak function of the gate voltage.

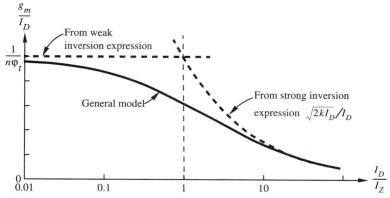

Figure 4.20 g_m/I_D from general model (solid line) compared to weak and strong inversion expressions (broken lines) [26, 27].

TABLE 4.6 A Saturation Region Model Valid in All Regions of Inversion*
(nMOS Transistor)

Drain current: $\quad I_D = I_Z \ln^2\left[1 + \exp\left(\dfrac{V_{GS} - V_T}{2n\phi_t}\right)\right]$ †

Small-signal conductances:

$$g_m = \frac{I_D}{n\phi_t}\,\frac{1}{f(x)}$$

$$g_{mb} = (n-1)g_m$$

$$g_{ds} = \frac{I_D}{V_A}$$

Small-signal intrinsic capacitances:

$$C_{gsi} = WLC'_{ox}\left[\frac{3}{2} + \frac{f(x)}{x}\right]^{-1}$$

$$C_{bsi} = WLC'_{ox}(n-1)\left[2 + \frac{f(x)}{x}\right]^{-1}$$

$$C_{gbi} = WLC'_{ox}\frac{n-1}{n}\left[1 - \left(\frac{3}{2} + \frac{f(x)}{x}\right)^{-1}\right]$$

$$C_{gdi} = C_{bdi} = 0$$

Noise power spectral densities:

$$S_{IW}(f) = 4\hat{k}T\left\{\frac{1}{2} + \frac{1}{6}\frac{x}{[f(x)]^2}\right\}ng_m$$

$$S_{If}(f) = \frac{hg_m^{\,2}}{WL}\cdot\frac{1}{f}$$

$$S_{VW}(f) = 4\hat{k}T\left\{\frac{1}{2} + \frac{1}{6}\frac{x}{[f(x)]^2}\right\}n\frac{1}{g_m}$$

$$S_{Vf}(f) = \frac{h}{WL}\cdot\frac{1}{f}$$

$$I_Z = 2Kn\phi_t^2$$

$$K = \frac{W}{L}K'$$

$$V_T = V_{T0} + \gamma(\sqrt{V_{SB} + \phi_0} - \sqrt{\phi_0})$$

$$n = 1 + \frac{\gamma}{2\sqrt{V_{SB} + \phi_0}} \ \ddagger$$

$$x = \frac{I_D}{I_Z}$$

$$f(x) = \sqrt{1 + 0.5\sqrt{x} + x}$$

---

* The current equation [29] has been adopted from Refs. 25 to 28 for using the source as a reference and for making possible the inclusion of an accurately evaluated V_T. The small-signal and noise expressions are from Ref. 27. The equation for C_{bsi} has been slightly modified for better accuracy in strong inversion. All symbols not defined here have been defined in Sec. 3.2, 4.2, 4.4, or 4.7, and $V_{DS} \geq 0$ is assumed. Junction leakage and breakdown effects are not included.

† If it is desired to include the dependence of I_D on V_{DS}, use $I_D = I'_D(1 + V_{DS}/V_A)$, where I'_D is the expression in the table.

‡ The value of n shown is a compromise for medium accuracy in all regions, and it is adequate for simple estimates. In weak inversion, better accuracy is obtained if n is replaced by m from Eq. (3.12). In strong inversion, if the expected voltage ranges are significant, better accuracy for I_D is obtained if n is replaced by α from Table 3.2. However, unless the transition between n, m, and α is made gradually as the various regions are entered, discontinuities will occur.

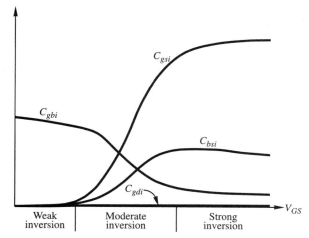

Figure 4.21 Intrinsic capacitances in saturation.

good example of how poor simulator models can be for analog applications. The modeling of small-signal capacitances in such models presents similar problems in the moderate inversion region, with errors of 200 percent in their values being possible. The implications of all this in circuit design are obvious.

How can such problems have gone unnoticed and uncorrected for so many years? The answer to this question lies in the fact that in the beginning integrated MOS transistors were used only in digital circuits, and the design needs of those circuits drove device modeling for a long time. In digital circuits the moderate inversion region is of minor importance, since the devices are usually either off or strongly on, only passing though moderate inversion briefly in the course of a transient. Thus models exhibiting the problems illustrated in Fig. 4.22 did not cause problems in digital design. In analog circuits, though, the moderate inversion region is important. For example, many devices in a low-voltage, low-power circuit can be operating in this region. Even in other circuits, some of the devices can be in moderate inversion; this is particularly true for the input devices in many operational amplifiers. In such cases, poor moderate inversion modeling can lead to poor design and to unnecessary design iterations.

The reader is urged to test any simulator he or she intends to use by generating plots like the one in Fig. 4.22*b,* using high resolution around suspected points such as $V_T + n\phi_t$ (Prob. 4.27). This is a very sensitive test of model performance, and its results are often shocking. Revealing such modeling problems will hopefully lead to efforts to correct them or will at least warn the designer not to trust the simulator

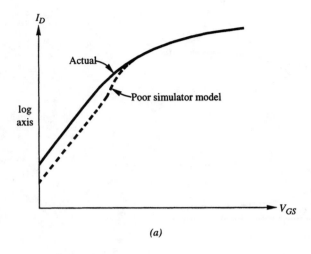

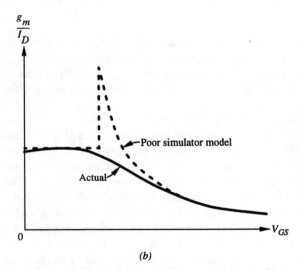

Figure 4.22 Comparison of actual characteristics to those obtained from common simulator models: (a) I_D (log axis) versus V_{GS}; (b) g_m/I_D versus V_{GS}.

and to "design around" such problems. A set of benchmark tests for evaluating models for analog design [30] is given in App. B.

4.9 Parameter Extraction for Accurate Small-Signal Modeling

In Sec. 3.12 we discussed the important process of model parameter extraction. We mentioned that model parameters are adjusted during this process until a certain error, between model prediction and measurements, is minimized. The selection of an appropriate error criterion depends on the application for which the model is intended. For analog applications, e.g., it is *not* enough to predict the drain current with an acceptable current error.* This has been discussed in Sec. 4.3, and is illustrated again in Fig. 4.23. In Fig. 4.23a, the solid line represents measured characteristics. The broken line represents a model whose parameters have been chosen so that I_D is predicted relatively accurately. Indeed, as shown, the error in predicting I_D is at most a few percent. In Fig. 4.23b, the *slopes* of the above two curves are shown—the slope predicted by the model is in error by a factor of 2! Since the slope of the I_D-V_{DS} characteristic is the small-signal drain conductance g_{ds}, the model is very poor for designing a variety of analog circuits. For example, the small-signal gain of a CMOS inverter is inversely proportional to the sum of the g_{ds} values of two devices, and can exhibit very large error if this model is used to calculate it. In such cases one needs a model that provides an accurate g_{ds}, which implies that the plot of I_D versus V_{DS} using such a model should track the nuances of experimentally observed behavior.

What compounds the problem with modeling g_{ds} is that in some CAD facilities, even if the model used is, in principle, capable of providing bearable accuracy for g_{ds}, the model parameters are chosen in such a way that this capability is not exploited. Thus, e.g., in some automated parameter extraction systems, model parameters are chosen so as to minimize the mean square relative error in the current. By this criterion the model in Fig. 4.23a would have been pronounced good, yet that model was seen in Fig. 4.23b to fail seriously in predicting g_{ds}. In cases where accurate g_{ds} prediction is of essence, a much better criterion for parameter extraction is the minimization of a combination of the mean square relative errors for I_D and for g_{ds}, appropriately weighted. With some models, use of the above criterion can provide a drastic improvement in g_{ds} accuracy, with a negligible loss in overall I_D accuracy. One must, of course, set up the appropriate facilities for measuring g_{ds}; this measurement can be tricky. Mean square errors involving g_m and g_{mb} could also be included in the criterion. However, these parameters, being significantly larger than g_{ds}, can be predicted relatively accurately if the current dependence on the bias voltages is predicted accurately.

In addition to errors in the saturation region considered above, significant errors can occur in the "transition" from nonsaturation to saturation. As remarked in Sec. 4.3.3, strong inversion models are not

---

* We are bringing up this point again, at the risk of being accused of being repetitive. This point appears to have been missed in the development of so many models that we would rather be repetitive than fail to give it the emphasis it deserves.

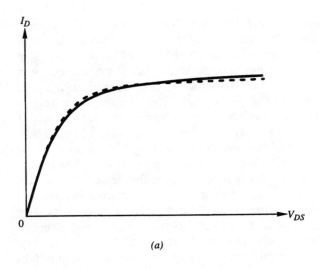

(a)

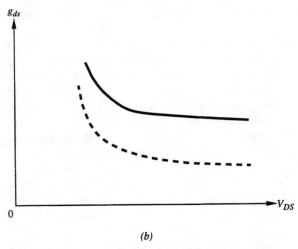

(b)

Figure 4.23 *(a)* I_D-V_{DS} characteristics; solid line is measured, broken line is the model. *(b)* Slope obtained from *(a)*.

accurate in this region since the drain end of the channel is then moderately inverted. Thus errors in g_{ds} can result.

4.10   Requirements for Good CAD Models

We conclude this chapter with a summary of requirements which a CAD model must ideally meet to be suitable for analog and mixed analog-digital circuit design [30].

1. The model should, of course, meet common requirements for digital work, such as reasonable *I-V* characteristic accuracy, digital logic speed prediction, etc.

2. It should give accurate values for all small-signal quantities such as $g_m, g_{mb},$ and g_{ds} and capacitances. In particular, all these parameters should be continuous with respect to any terminal voltage.

3. It should give good results even when the device operates non-quasistatically, or at least it should degrade gracefully for such operation, as frequency is increased.

4. It should give accurate predictions for both white and $1/f$ noise, including the triode region.

5. It should meet requirements 1 to 4 above over large bias ranges, including $V_{SB} \neq 0$, and encompassing the weak, moderate, and strong inversion regions.

6. It should do all the above over the temperature range of interest.

7. It should do all the above for any combination of channel width and length values, from the minimum specified upward.

8. The user should only have to specify the geometric dimensions for each device and one set of model parameters, valid for all devices of the same type and independent of dimensions.

9. The model should provide a flag every time there is an attempt to use it outside its limits of validity. For example, if the model is quasistatic and one attempts to use it, say, around the unity-gain frequency of the device, the user should be warned that the result may be inaccurate.

10. The model should have as few parameters as possible (but just enough), and those parameters should be linked as strongly as possible to ones related to the device structure and fabrication processing (e.g., oxide thickness, substrate doping, junction depth). This would allow meaningful worst-case simulations and predictions. Empirical parameters without physical meaning should be avoided as much as possible. This requirement strongly points in the direction of a physics-based model.

11. The model should be linked to an efficient parameter extraction method. The number of required test devices and tests for parameter extraction should be as small as possible.

12. The model should ideally provide links to device and reliability simulators.

To the author's knowledge, at the time of this writing no model exists which satisfies all the above requirements, and in fact most models fail most of them. MOSFET modeling for CAD still has a long way to go. Readers may want to test the models they use with the benchmark tests given in App. B.

References

1. Y. P. Tsividis, *Operation and Modeling of the MOS Transistor,* McGraw-Hill, New York, 1987.
2. H. C. de Graaff and F. M. Klaassen, *Compact Transistor Modeling for Circuit Design,* Springer-Verlag, New York, 1990.
3. N. Arora, *MOSFET Models for VLSI Circuit Simulation—Theory and Practice,* Springer-Verlag, New York, 1993.
4. K. Lee, M. Shur, T. A. Fjeldly, and T. Ytterdal, *Semiconductor Device Modeling for VLSI,* Prentice-Hall, Englewood Cliffs, NJ, 1993.
5. G. Merckel, J. Borel, and N. Z. Cupcea, "An Accurate Large-Signal MOS Transistor Model for Use in Computer-Aided Design," *IEEE Transactions on Electron Devices,* 19: 681–690, May 1972.
6. P. Antognetti and G. Massobrio, *Semiconductor Device Modeling with Spice,* McGraw-Hill, New York, 1993.
7. G. T. Wright, "Physical and CAD Models for the Implanted-Channel VLSI MOS-FET," *IEEE Transactions on Electron Devices,* 34: 823–833, April 1987.
8. A. R. Boothroyd, S. W. Tarasewicz, and C. Slaby, "MISNAN-A Physically Based Continuous MOSFET Model for CAD Applications," *IEEE Transactions on Computer-Aided Design,* 10: 1512–1529, December 1991.
9. C. McAndrew, B. K. Bhattacharyya, and O. Wing, "A Single-Piece Continuous MOS-FET Model Including Subthreshold Conduction," *IEEE Electron Device Letters,* 12: 565–567, 1991.
10. J. A. Power and W. A. Lane, "An Enhanced Spice MOSFET Model for Analog Applications," *IEEE Transactions on Computer-Aided Design,* 11: 1418–1425, November 1992.
11. N. D. Arora, R. Rios, C.-L. Huang, and K. Raol, "PCIM: A Physically Based Continuous Short-Channel IGFET Model for Circuit Simulation," *IEEE Transactions on Electron Devices,* 41: 988–997, June 1994.
12. J. H. Huang, Z. H. Liu, M. C. Jeng, P. K. Ko, and C. Hu, "A Physical Model for MOS-FET Output Resistance," *Digest of Technical Papers, International Electron Devices Meeting,* San Francisco, CA, 1992.
13. R. M. D. A. Velghe, D. B. M. Klaassen, and F. M. Klaassen, "Compact MOS Modeling for Analog Circuit Simulation," *Digest of Technical Papers, International Electron Devices Meeting,* Washington, DC, 1993, pp. 485–488.
14. T. Skotnicki, C. Denat, P. Senn, G. Merckel, and B. Hennien, "MASTAR: A New Analog/Digital CAD Model for Sub-halfmicron MOSFETs," *Proceedings of the International Electron Devices Meeting,* San Francisco, CA, 1994.
15. S. M. Sze, *Physics of Semiconductor Devices,* Wiley, New York, 1981.
16. L. O. Chua, C. A. Desoer, and E. S. Kuh, *Linear and Nonlinear Circuits,* McGraw-Hill, New York, 1987.
17. D. H. Treleaven and F. N. Trofimenkoff, "MOSFET Equivalent Circuit at Pinchoff," *Proceedings of the IEEE,* 54: 1223–1224, 1966.
18. W. H. Hayt, Jr., and J. E. Kemmerly, *Engineering Circuit Analysis,* McGraw-Hill, New York, 1993.
19. M. Bagheri and Y. Tsividis, "A Small-Signal DC to High Frequency Nonquasi-static Model for the Four Terminal MOSFET Valid in All Regions of Operation," *IEEE Transactions on Electron Devices,* 32: 2383–2391, November 1985.
20. K.-W. Chai and J. J. Paulos, "Unified Nonquasi-static Modeling of the Long-Channel Four-Terminal MOSFET for Large and Small-Signal Analyses in all Operating Regimes," *IEEE Transactions on Electron Devices,* 36: 2513–2520, November 1989.
21. P. Roblin, S. C. Kang, and W.-R. Liou, "Improved Small-Signal Equivalent Circuit Model and Large-Signal State Equations for the MOSFET/MODFET Wave Equation," *IEEE Transactions on Electron Devices,* 38: 1706–1718, August 1991.

22. A. Papoulis, *Probability, Random Variables, and Stochastic Processes,* McGraw-Hill, New York, 1965.
23. A. Van der Ziel, *Noise: Sources, Characterization, and Measurement,* Prentice-Hall, Englewood Cliffs, NJ, 1970.
24. A. Ambrozy, *Electronic Noise,* McGraw-Hill, New York, 1982.
25. H. J. Oguey and S. Cserveny, "MOS Modeling at Low Current Density," Summer Course on Process and Device Modeling, K U Leuven, 1983.
26. E. A. Vittoz, "Micropower Techniques," in J. E. Franca and Y. Tsividis, eds., *Design of Analog-Digital VLSI Circuits for Telecommunications and Signal Processing,* Prentice-Hall, Englewood Cliffs, NJ, 1994.
27. C. Enz, *High Precision CMOS Micropower Amplifiers,* Ph.D. thesis no. 802, Ecole Polytechnique Federal de Lausanne, Switzerland, 1989.
28. C. C. Enz, F. Krummenacher, and E. A. Vittoz, "An Analytical MOS Transistor Model Valid in All Regions of Operation and Dedicated to Low-Voltage and Low-Current Applications," *Analog Integrated Circuits and Signal Processing,* 8:83–114, July 1995.
29. Y. Tsividis, K. Suyama, and K. Vavelidis, "A Simple 'Reconciliation' MOSFET Model Valid in all Regions," *Electronics Letters,* 31: no. 6, 506–508, March 1995.
30. Y. Tsividis and K. Suyama, "MOSFET Modeling for Analog Circuit CAD: Problems and Prospects," *IEEE Journal of Solid-State Circuits,* 29: 210–216, March 1994.

Problems

Note: All transistors are assumed to be nMOS at room temperature (300 K), unless noted otherwise. Mobility dependence on bias and short- and narrow-channel effects are to be neglected, unless stated otherwise.

4.1 Equation (4.9) gives three equivalent expressions for g_m in strong inversion saturation. Assume that $V_{GS} = V_T$ and $I_D = I_{DX}$. Then the three expressions predict g_m values of 0, $\sqrt{2kI_{DX}}$, and infinity. How is this possible, if the expressions are supposed to be equivalent?

4.2 A device characterized by the parameters of Example 3.8 is biased in saturation with a current of 140 µA. If $L = 3$ µm, find the value of W needed to provide a transconductance of (a) 360 µA/V and (b) 4.5 µA/V. Assume $V_{SB} = 0$ V. Make sure you check your assumptions about the region of inversion in which the device is operating for each case.

4.3 Plot the ratio g_{mb}/g_m versus V_{SB}, with V_{SB} from 0 to 5 V, and γ as a parameter with values from 0.25 to 1 $V^{1/2}$, in steps of 0.25 $V^{1/2}$. Use a single value of $\phi_0 = 0.8$ V for simplicity. Assume operation in saturation and the following two cases:

(a) Operation in weak inversion
(b) Operation in strong inversion with $V'_{DS} = 2$ V

4.4 A device characterized by the parameters of Example 3.8 is biased in saturation with a current of 200 µA and $V_{SB} = 2$ V. Find the values of W and L needed to provide a g_m of 400 µA/V and a g_{ds} of 4 µA/V. What is the resulting value of g_{mb}?

4.5 Show that *under certain conditions* (which you should state), g_m in saturation is approximately equal to g_{ds} at $V_{DS} = 0$, if one assumes the same value of V_{GS} and V_{SB}. Show that in contrast to what is sometimes assumed, the two quantities are, in general, different.

4.6 For a device with a given W and L, give qualitative plots of g_m, g_{ds}, and g_m/g_{ds} versus I_D, with I_D varying from weak to strong inversion. In moderate inversion you can use a guess, simply joining the weak and strong inversion segments with a smooth curve. (Such guessing will not be necessary after Sec. 4.8 is covered; see Prob. 4.25.)

4.7 Can the "back-gate" transconductance g_{mb} become as large as the "front-gate" transconductance g_m? If so, under what conditions?

4.8 (a) Derive an expression for the body transconductance g_{mb} in the form bg_m, where b is an appropriate expression, starting from the strong inversion nonsaturation symmetric model in Eq. (3.5). What is the drawback of this expression? (*Hint:* Consider V_{DS} approaching 0 and the numerical problems that can arise if this expression is used unmodified as part of a computer simulation model.)
(b) Assuming the above expression is very accurate, calculate the worst-case error of our expression for the strong inversion nonsaturation g_{mb} in Table 4.1, assuming V_{SB} and $V_{DB} = V_{DS} + V_{SB}$ can be anywhere between 0 and 5 V. Assume $\phi_0 = 0.76$ V.
(c) An often used expression for the strong inversion nonsaturation g_{mb} is $g_m[\gamma/(2\sqrt{V_{SB} + \phi_0})]$. Find the worst-case error of this expression for the conditions in (b).

4.9 Find an expression for g_m in strong inversion saturation in the presence of effective mobility dependence on V_{GS} (Sec. 3.6).

4.10 Using the results of Prob. 3.25, show that if strong velocity saturation is present, g_m is practically independent of V_{GS} and L.

4.11 Determine the complete small-signal equivalent circuit for a transistor with $W = 400$ μm, $L = 4$ μm, and 400-μm by 3-μm rectangular source and drain regions. Assume parameters as in Example 3.8, $V_{SB} = 1$ V, $V_{DS} = 4$ V, and
(a) $V_{GS} = 2$ V
(b) $V_{GS} = 0.85$ V

4.12 Repeat Prob. 4.11 for $W = 4$ μm, $L = 400$ μm, and 4-μm by 3-μm rectangular source and drain regions. Comment on the relative importance of certain capacitances here (where $W/L \ll 1$) and in Prob. 4.11 (where $W/L \gg 1$).

4.13 It is desired to develop a current of 50 μA by biasing an *n*MOS device with an appropriate V_{GS}. Assume $V_{SB} = 0$ V and $V_{DS} = 3$ V. Find the total gate-drain and body-drain capacitances for the following two cases (a different W/L must be chosen in each):

(a) Operation in weak inversion, with $I_D = I_M/4$
(b) Operation in strong inversion, with $V_{GS} - V_T = 0.5$ V

4.14 For the device in Example 4.6, calculate (a) the intrinsic cutoff frequency and (b) the cutoff frequency including extrinsic parasitic effects.

4.15 Prove the expression for the weak inversion value of f_{Ti} given in Eq. (4.18).

4.16 In the circuit of Fig. 4.12a, assume that the transistor has parameters as in Example 3.8, $W = 10$ μm, $L = 5$ μm, $V_{GS} = 2$ V, $V_{SB} = 0$ V, and $V_{DS} = 3$ V. Consider *only* the *intrinsic* part of the device. The small-signal voltage is $\epsilon \sin \omega t$, as shown. Let the magnitude and phase of the input small-signal current be ϵ_i and θ_i, respectively, and let the corresponding quantities of the output small-signal current be ϵ_o and θ_o, respectively. Plot the quantities ϵ_o/ϵ_i, ϵ_o/ϵ, θ_i, θ_o, and $\theta_o - \theta_i$ versus frequency, using a logarithmic frequency axis and frequency values from $0.0001 f_{Ti}$ to f_{Ti}, where f_{Ti} is the intrinsic cutoff frequency of the device (Sec. 4.5), for the following two cases:

(a) Use the small-signal model of Fig. 4.7, which does not include nonquasistatic effects.
(b) Use the small-signal model of Fig. 4.13, which includes first-order nonquasistatic effects.

Compare and comment.

4.17 Show that the nonsaturation formulas for the noise power spectral density in Table 4.4 reduce to the saturation ones at $V_{DS} = V'_{DS}$.

4.18 For a transistor with $W = 100$ μm and $L = 6$ μm, the equivalent input noise voltage contributed by $1/f$ noise in the band 1 Hz to 1 MHz is 17 μV. Find the value of the constant h in Tables 4.4 and 4.5.

4.19 Consider a pMOS transistor in saturation with $W = 3$ μm, $L = 30$ μm, $V_{SB} = 0$, and $I_D = 10$ nA. Find the $1/f$ and white noise contributions to the drain current noise, as well as the combined noise, in the frequency bands
(a) 1 to 20 Hz
(b) 1 to 20 MHz
Use the parameter values of Example 3.11, and $h = 8 \times 10^{-11}$ $V^2 \cdot \mu m^2$.

4.20 Show that the total $1/f$ noise contribution over a band of frequencies f_1 to f_2 depends on the ratio f_2/f_1 (as opposed to the individual values of f_1 and f_2).

4.21 A designer must choose W and I_D for a transistor with a given L, so that its gate transconductance becomes equal to a given value. It is desired to keep the transistor's equivalent input noise voltage small. Is it preferable to bias the device in weak or in strong inversion? Answer the above question if the main noise of concern is
(a) $1/f$ noise
(b) White noise

4.22 An expression found in popular simulators for the power spectral density of the drain current thermal noise in strong inversion is $4\hat{k}T(2g_m/3)$, where the meaning of the symbols is as in Sec. 4.7. Based on the results given in that section:

(*a*) Show that the above expression is totally wrong deep in nonsaturation and that, in fact, it predicts that the channel is noiseless at $V_{DS} = 0$.

(*b*) Show that the correct expression for the above quantity at $V_{DS} = 0$ is equivalent to $4\hat{k}TR$, where R is the channel resistance.

4.23 For the device of Examples 4.1 and 4.2, V_{GS} is changed until the device is in moderate inversion with $I_D = I_Z$ (see Fig. 3.9 and Table 4.6). Determine the values for all elements in its small-signal equivalent circuit (Fig. 4.7).

4.24 Assuming the general model of Sec. 4.8 is accurate, calculate the errors in the weak and strong inversion approximations at the weak inversion upper limit and at the strong inversion lower limit (as given in Tables 3.1 and 3.2), respectively, for the following quantities: (*a*) I_D, (*b*) g_m, (*c*) g_m/I_D, (*d*) C_{gsi}, and (*e*) C_{gbi}.

4.25 Repeat Prob. 4.6, using the results of Sec. 4.8.

4.26 A transistor with a given L is to be biased in saturation with a dc voltage V_{GS} so that it produces a given bias current value. Consider the noise in the resulting current. What are the tradeoffs between choosing (*a*) a small W with a large V_{GS} and (*b*) a large W with a small V_{GS}? Specifically comment on how the white and $1/f$ noise components vary with the choice of W (for the given L and I_D). Your investigation should include weak, moderate, and strong inversion.

4.27 Use a computer simulator to plot I_D, g_m, and g_m/I_D versus V_{GS} from weak to strong inversion in saturation. Use a fine resolution around $V_{GS} = V_T + n\phi_t$, and check whether the models used in your simulator have the problems discussed in Sec. 4.8. If you are using the simulator Spice2G or equivalent, use the nMOS parameters given in App. C. With $W = L = 10$ μm, $V_{SB} = 0$ V, $V_{DS} = 4$ V, and V_{GS} values between 0.7 and 0.9 V (1-mV resolution), the problem should be clearly visible slightly above 0.8 V. If not, check over a wider range.

4.28 Run all benchmark tests given in App. B on the model(s) at your disposal.

4.29 Judge the model(s) at your disposal against the requirements listed in Sec. 4.10. (This problem can be done best if Prob. 4.28 is done first.)

Technology and Available Circuit Components

5.1 Introduction

Device-level circuit design in VLSI requires a rather intimate knowledge of how devices are made. This is in contrast to the design of circuits made with discrete components, where one can use a commercially available transistor, knowing only its electrical parameters but not what it looks like inside the package or how it was made. In VLSI, a knowledge of device structures affords two advantages to the circuit designer: (1) The designer can choose device shapes, dimensions, and placement (the latter as part of the "layout" processes, discussed in Chap. 6); this represents a degree of freedom without which many state-of-the-art circuits would not have been possible. (2) The designer is able to determine the ways in which different devices, packed closely together on one chip, can interfere with each other and to devise ways to eliminate such interference.

This chapter is devoted to fabrication technology (often referred to simply as *technology* or *fabrication processing* or simply *processing*). A detailed exposition of fabrication technology can be found elsewhere [1–6]. The information provided here can be considered as the minimum needed for the design of relevant, working circuits (as opposed to circuits that work only "in principle," i.e., on paper). We will consider the most important fabrication processes and will describe the components such processes make available to the designer, including transistors, resistors, and capacitors. The reader is certainly familiar with the ideal models (i.e., the defining relations) of isolated resistors and capacitors. In this chapter we will provide additional information that is necessary to model properly these components *in their environment* (the chip of which they are part), including the unavoidable parasitic elements. Integrated inductors, which are being used in very high-frequency circuits, will be considered, too. We will also discuss the modeling of bipolar transistors; however, since we have already presented the modeling process in detail while using the MOS transistor as an example, here we will be rather brief.

The first fabrication process used for making commercial chips was the *bipolar* process, called so because it is used for making bipolar transistors. Bipolar processes have continued to evolve to the present day, achieving remarkable speed performance. However, when it comes to VLSI, circuit size and power considerations limit their use. Thus, most VLSI chips are made by using MOS technology. The first fabrication process used for making MOS chips was the *PMOS* process, so named because all transistors on the chip were pMOS devices. After a number of technological advances, this process was gradually replaced by the *NMOS* process, in which all transistors on the chip were nMOS devices; the higher mobility of electrons, compared to that of holes, made higher operating speeds possible. At about the same time, the *CMOS* (complementary MOS) process was developed, in which both nMOS and pMOS devices coexist on the same chip. This process provided extra flexibility to circuit design, and made possible reduced power dissipation for digital circuits; it is now the dominant VLSI process. A more recent VLSI process is the *BiCMOS* (bipolar CMOS) process, which in addition to nMOS and pMOS devices makes available high-performance bipolar transistors. The emphasis in this chapter is on CMOS and BiCMOS processes. Chip fabrication will be illustrated by using an industry workhorse as an example—the n-well CMOS process.

5.2 The n-Well CMOS Process

5.2.1 Basic fabrication steps and MOS transistor structures

Before discussing the basic fabrication steps of a simple CMOS process, we take a look at the final result of such steps. MOS transistors made with a CMOS process are shown in Fig. 5.1a and b. In Fig. 5.1a, a top view is shown. If one makes a cut along a horizontal line passing through the middle, one obtains the vertical cross section shown in Fig. 5.1b. In this figure and in other related figures in this chapter, the approximate boundary between silicon regions of high and low doping of the same conductivity type (i.e., both p or both n), is indicated by broken lines. Boundaries between regions of the opposite type (which thus form a pn junction) and boundaries between different materials are indicated by solid lines. The substrate is p type (usually with a doping concentration between 10^{14} and 10^{16} cm^{-3}), and the nMOS devices can be formed directly on it. However, pMOS devices must have an n-type body and thus cannot be formed directly on the substrate. Thus, separate n-type regions called *wells* or *tubs* are used to host such devices as shown. For this reason, this particular process is referred to as an *n-well process*. The body of the pMOS transistors is contacted through an n^+ region in the n well, to enable a good ohmic contact (as opposed to a rectifying contact) to be made. The body of the nMOS transistor is contacted through p^+ regions as shown; the back side of the wafer can also be used for such contacts.

Both nMOS and pMOS devices are enhancement type in standard CMOS processes, and ion implantation is used to define their threshold voltage. The most common gate material is polycrystalline silicon, known as *polysilicon* or *poly*. The resulting process is called a (*poly*)-*silicon-gate process*.

In Fig. 5.1b, the gates are shown embedded in insulator material, vaguely referred to as *oxide*. This oxide is "thin" under the gate (60 to 200 Å in modern processes*) and "thick" elsewhere (typically 3000 to 10,000 Å). Metal is used to provide contacts through *contact windows*, which are holes cut through the oxide to allow the metal to contact the desired region.

The substrate is common for all nMOS devices on the chip and thus is *not* available as a terminal for individual nMOS devices. In some circuits, voltages can vary between a positive power supply voltage $(+V_{DD})$ and a negative power supply voltage $(-V_{SS})$, both taken with respect to a ground potential. In other circuits, a single power supply $(+V_{DD})$ is used; for such cases $-V_{SS}$ in our figures can be assumed to represent the ground potential. To make sure that the pn junctions formed by the p substrate and the various n regions are not forward-biased, the substrate is connected to the most negative voltage in the circuit $(-V_{SS})$. Thus, the common body of all nMOS devices is permanently attached to $-V_{SS}$, as shown in Fig. 5.1c. In contrast to this, all four terminals of pMOS devices can be available as shown, provided each device is placed in its own n well. If in a given circuit several pMOS devices can have a common substrate connection, and provided they will not interfere with each other (Chap. 6), they can all share a common n well, thus saving chip area. In either case, the well-substrate pn junction is permanently connected between the pMOS transistor body and $-V_{SS}$, as shown. This junction is reverse-biased, but its capacitance (and sometimes its leakage current) must be taken into account in circuit design. In Fig. 5.1c we show it in broken lines because in circuit diagrams it is usually omitted for simplicity, although its presence should be kept in mind.

---

* $1 \text{ Å} = 10^{-4} \text{ μm} = 10^{-10} \text{ m}$.

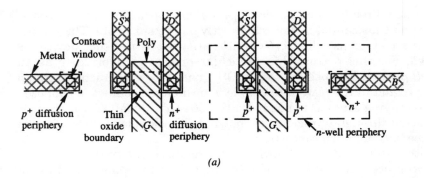

(a)

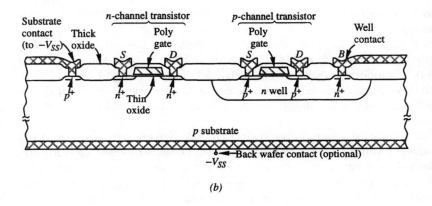

(b)

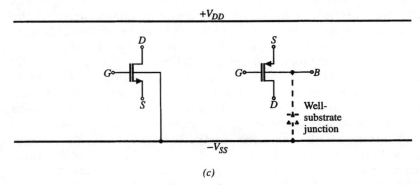

(c)

Figure 5.1 Transistors in an n-well CMOS process. (*a*) Top view; (*b*) cross section along a horizontal line passing through the middle of *a*; (*c*) power-supply connection and available device terminals (the diode shown models the well-substrate junction).

The fabrication of MOS integrated circuits (ICs) starts with a silicon circular wafer, about 0.5 mm thick and usually with a diameter of 15 or 20 cm (about 6 to 8 in). Usually, many chips (identical or different) can be made on each wafer.* Like features of all like devices on the wafer are fabricated simultaneously. Thus, the fabrication cost per wafer is shared among the chips on it. IC fabrication involves several different steps, such as lithography, etching, and implantation; some fabrication steps can be performed on an entire wafer lot (25 to 50 wafers) simultaneously, which decreases costs. The shapes and locations of each type of feature on the chip are transferred onto the wafer from *masks* (so named because in their simplest form they are used to selectively mask light during a photolithographic process, through which patterns are transferred onto the wafer). Each mask contains the shapes and locations of one *type* of feature, but for all like features on the wafer. For example, a poly mask can indicate all areas where polysilicon is to eventually exist, for all devices on all chips on the wafer.

IC fabrication processing has evolved into a truly complex art [1–4], with numerous details which are not within the scope of this book. Here we will only provide a flavor by sketching some of the major steps in a typical n-well CMOS process (resulting in the structures of Fig. 5.1), with the help of Fig. 5.2. In each part of this figure (a through g), the upper half represents a small part of a mask used in the fabrication process. The lower half shows a cross section of the corresponding part of the chip, corresponding to a horizontal line passing through the center of the mask in the upper half.

In Fig. 5.2a, the p-type substrate has been covered with oxide, in which openings have been etched where wells are to be made, as defined on the mask. The wafer is implanted with n-type impurities. The implant ions reach the surface of the wafer through the openings, as shown. During a high-temperature fabrication step, these implanted impurities diffuse downward (typically by 2 to 6 μm), forming the n well. The boundary of the well after this diffusion process is shown in the figure. The impurities also diffuse sideways, a phenomenon referred to as *lateral diffusion;* the extent of the lateral diffusion is typically 0.7 times the vertical extent. Thus, the size of the well (as seen from above) ends up being larger than the corresponding area on the mask; it is important to keep this in mind.

The following steps are to define the *active areas* (areas where n^+ regions, p^+ regions, or transistor channels are to be formed) and to "grow" the gate oxide. At the end of these steps there is thin oxide (typically 60 to 200 Å in modern process) over such areas and thick oxide

---

* A rare exception is *wafer-scale integration,* in which the "chip" occupies the entire wafer [7].

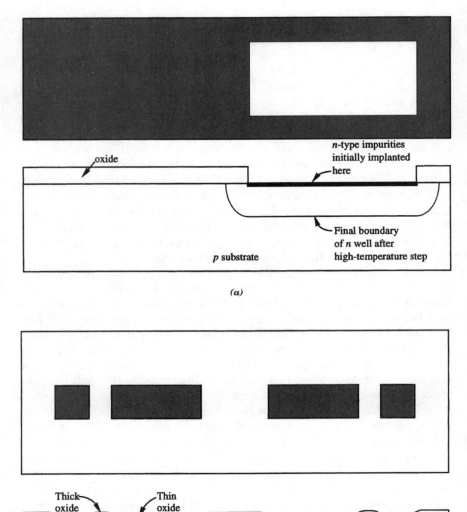

Figure 5.2 Basic fabrication steps. In each part, the top half shows a portion of a mask; the bottom half shows the corresponding cross section of the chip corresponding to a horizontal line passing through the middle of the top part. The parts above show the formation of (a) wells, (b) active areas.

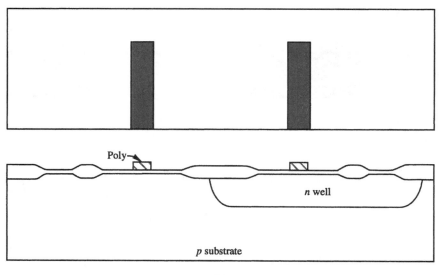

(c)

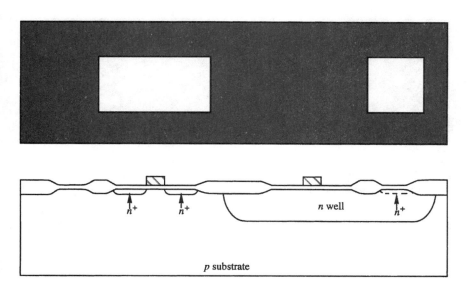

(d)

Figure 5.2 *(Continued)* Basic fabrication steps. In each part, the top half shows a portion of a mask; the bottom half shows the corresponding cross section of the chip corresponding to a horizontal line passing through the middle of the top part. The parts above show the formation of *(c)* poly regions, *(d)* n^+ implant regions.

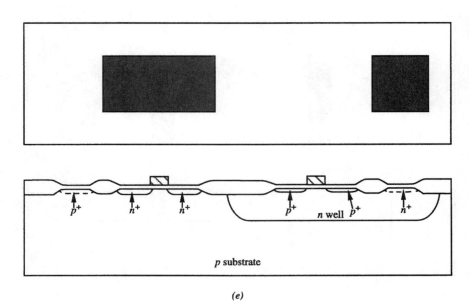

(e)

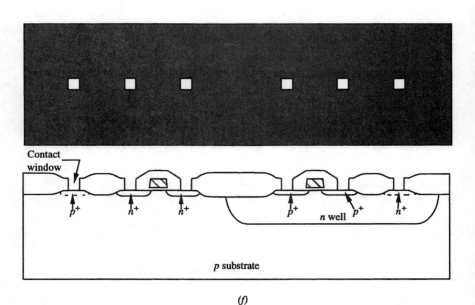

(f)

Figure 5.2 (*Continued*) Basic fabrication steps. In each part, the top half shows a portion of a mask; the bottom half shows the corresponding cross section of the chip corresponding to a horizontal line passing through the middle of the top part. The parts above show the formation of (e) p^+ implant regions, (f) contact windows.

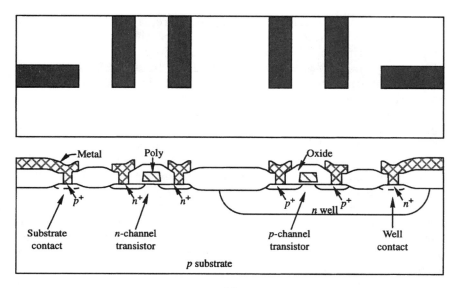

(g)

Figure 5.2 (*Continued*) Basic fabrication steps. In each part, the top half shows a portion of a mask; the bottom half shows the corresponding cross section of the chip corresponding to a horizontal line passing through the middle of the top part. The parts above show the formation of metal regions.

everywhere else (referred to as *field oxide*), as shown in Fig. 5.2b. Oxide grows partly upward and partly downward from the original silicon surface. The gradual transition from thick to thin causes an encroachment of the field oxide into the active areas and is referred to as *bird's beak*.

The next step is to cover the wafer with poly, dope the poly heavily with n-type impurities, and etch (remove) the poly except in areas defined by a mask, as shown in Fig. 5.2c. For the particular example we are using, these poly regions will form the transistor gates. At other points on the wafer, poly strips can also be left unetched to form interconnection wires or resistors; such strips, though, in contrast to the gate poly shown, will be over thick oxide. Due to fabrication idiosyncracies [1–4], the width of the poly layers ends up somewhat different from the corresponding dimension on the mask. This is also true of other layers.

Next, n-type impurities are implanted in regions indicated by another mask, shown in the top half of Fig. 5.2d. Within such regions only the *active* areas, which are covered by only thin oxide, let the implant go through, and then again only if there is no poly in the way. Thus, the n^+ impurities end up in the areas shown in the bottom half of

Fig. 5.2*d*. As seen in the left part of the figure, the poly prevents the impurities from landing under it, and hence the source and drain regions are formed on either side of the gate, just outside its "shadow." This is why the poly-gate process is referred to as a *self-aligned gate* process. Note that the poly receives the full implant. The outer boundary of the n^+ regions is somewhat different from that defined by the active area mask, partly due to the bird's beak effect mentioned above. For simplicity, in drawing the n^+ regions we have anticipated a high-temperature step, which occurs later in the process. In this step, the n^+ impurities diffuse downward but also laterally, so they creep under the gate somewhat; nevertheless, the resulting overlap distance between the gate and the source/drain regions remains small (e.g., 0.1 to 0.2 μm), which is desirable since it results in small gate-drain and gate-source overlap parasitic capacitances.* The n^+ regions formed in this step can be used as sources or drains for the nMOS transistors, as contacts to the n wells, and in some cases as resistors and parts of the chip wiring interconnects.

Next, the complement of the previous mask is used, to indicate where *p*-type impurities are to be implanted. This is shown in Fig. 5.2*e*. Here the remaining active areas, which were prevented from being implanted in the previous step, are exposed. Again, within the exposed areas only the thin oxide of the active areas lets the implant go through, and then only if there is no poly in the way. This step forms the source or drain regions of the *p*MOS transistors, the contacts to the *p* substrate, and in some cases resistors and parts of the wiring. (Recall that the poly was doped n type immediately before this step. Now that it receives *p*-type impurities, its effective doping concentration changes, but still remains n type because the initial n-type doping is very heavy. Following a high-temperature step, the p^+ regions are as shown. As before, note the effect of lateral diffusion under the gate and the change in the size of the p^+ strips due to the bird's beak effect. The n^+ and p^+ regions are sometimes jointly referred to as *diffusion* regions.†

An oxide is then deposited, and contact windows are opened through it, as indicated by a corresponding mask. This is shown in Fig. 5.2*f*. The wafer is then covered by metal, which is subsequently selectively removed except in areas specified by a mask, as shown in Fig. 5.2*g*. The metal that remains can be used to contact the n^+, p^+, and poly regions

---

* This was not true for the old metal-gate processes, in which the source and drain regions were formed before the gate using a separate mask, and the metal gate had to overlap those regions to a significant extent to ensure coverage in the presence of mask misalignment.

† This name originated in the days before CMOS, when wells (which are also diffused) did not exist and confusion could not arise.

(contacts to the latter are not shown in the figure) and is the main layer used for interconnecting the various devices in the circuit. The interconnecting metal runs over thick oxide.

The above description of basic fabrication steps represents the minimum amount of information needed to relate masks to the various features of finished devices. This minimum was our goal, and we have avoided getting into details. The reader should be aware, though, that the details we have skipped are many. Each step we have sketched above actually consists of several involved substeps [1–4]. In addition, several other steps are used in between the steps we have described. For example, extra ion implants are used to adjust the threshold voltage of the transistors or to reduce punch-through effects. Also, an implant is used before the field oxide is grown, to create the channel-stop areas briefly mentioned in Sec. 4.4.1; the reason for using such areas is the following. Metal and poly are used to interconnect the various devices on the chip; such interconnecting "wires" run over thick oxide. There exists the danger that these layers can act as undesired "gates" and create a parasitic channel in the substrate below them, which then would act as a conductive path between devices. To prevent this, one must ensure that the parasitic device formed by the interconnection layer, the thick oxide, and the substrate has a sufficiently high threshold voltage (referred to as the *field* threshold). This is achieved by increasing the p-type doping concentration of the silicon surface outside the nMOS transistors, using a *field implant*. Since the resulting regions prevent the parasitic channels from forming, they are called *channel-stop* regions; they are not shown in the figures of this section for simplicity. The channel-stop regions also help avoid punch-through between closely located source and drain regions of different transistors. The presence of the channel-stop regions affects the effective location of the boundary of the n^+ and p^+ regions, and is one more reason that the widths of such regions can be somewhat different from that on the mask.

Two layers of metal (separated by thick oxide) are used for extra wiring flexibility. In double-metal processes, the second metal layer is separated from the first by thick oxide, and where desired, the two layers can be brought into contact through metallized openings in the oxide called *vias*. Three, four, or five layers of metal, for even higher wiring flexibility, are sometimes used, each being separated from the one below it by thick oxide. Two layers of poly, referred to as *poly 1* and *poly 2,* are sometimes used (resulting in the *double-poly* processes); these, separated by a thin oxide layer, are used to create high-quality capacitors (Sec. 5.2.2). Alternatively, capacitors can be created between a metal layer and a poly layer. Extra fabrication steps are also used to lower the resistivity of n^+, p^+, and poly regions by forming low-resistance *silicides* on them [1–6]. All these variations add to the fabrication cost. While a simple CMOS process can involve 10 masks, more sophisticated ones can use more than 20. A full description of fabrication processes can involve well over a hundred detailed steps.

After metallization the wafer is coated with a protective glass layer, and openings are cut through it over the *pads,* which are metal areas needed for connecting the chip to the external world. Of these pads, some will be used for permanent connection; these *bonding pads* are large (typically 100 µm by 100 µm) and are placed by the designer near the periphery of each chip. Other pads are used only for temporary testing through probing, and they can be smaller.

Following fabrication, the circuits are often tested on each wafer by using special probes, and the units that do not function properly are marked. The wafer is subsequently cut into individual chips, and the good ones are attached to packages. In a popular packaging technique, the bonding pads are connected to the package leads through bonding wires; contact is also often made to the back of the chip. The packages are then sealed. The packaged chips are tested again, to discard the ones that may have suffered damage in the steps following wafer probe testing, and in some cases to make sure that the chips function properly in the presence of effects related to packaging (e.g., package parasitic capacitance or mechanical strain).

In addition to nMOS and pMOS transistors, a number of other devices can be available in a CMOS process. These will now be discussed.

5.2.2 Capacitors

A top view of a parallel-plate capacitor is shown in Fig. 5.3a. The value of the capacitance, excluding parasitics, is given by

$$C = AC' \tag{5.1}$$

where A is the total area of the top plate over thin oxide and C' is the capacitance per unit area. The insulator of capacitors in common CMOS processes is usually thicker than the transistor gate oxide. This improves matching (Sec. 5.7), since random thickness fluctuations become a smaller fraction of the total thickness. Typical values for C' are 0.6 to 1.2 fF/µm^2. One should keep in mind that extra capacitance due to interconnections and even due to the sidewalls along the periphery of the capacitor plates will be present; this is not included in the above formula.

Several ways of making capacitors are used in CMOS technology [8]. A popular one is shown in Fig. 5.3b. The bottom plate is made of polysilicon. In double-poly technology, which makes available two poly levels, the top plate can also be made of polysilicon. Otherwise, metal

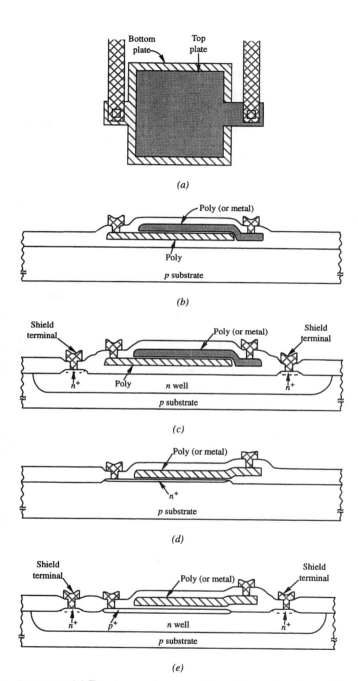

Figure 5.3 (*a*) Top view; and cross sections of capacitors in *n*-well CMOS technology: (*b*) poly-poly (or metal-poly) capacitor; (*c*) the capacitor of *b* with a well as bottom shield; (*d*) poly-n^+ or metal-n^+ capacitor; (*e*) poly-p^+ or metal-p^+ capacitor with a well as bottom shield.

is used for the top plate. A high-quality thin oxide must be formed as the insulator, before the top plate is formed.

Two parasitic capacitors are associated with the main capacitor. The main one is between the bottom plate and the substrate. The value of this capacitor can typically be about 10 percent of the main capacitance. This value is augmented by the extra capacitance of the metal line used to contact the bottom plate. The total capacitance due to the above effects is an unavoidable parasitic, permanently attached to the main capacitor, as shown in Fig. 5.4. In addition to this, the metal wiring used to contact the top plate results in a small parasitic capacitance to the substrate. This capacitance, also shown in Fig. 5.4, depends on the wiring details. It is usually at least several femto-farads. Another parasitic, shown in Fig. 5.4 in broken lines, is the resistance of the polysilicon plate(s). This parasitic way may be ignored except at very high frequencies, at which the reactance of the capacitor [9] is no longer much larger than the parasitic resistance.

The capacitor of Fig. 5.3b is susceptible to interference. Any "noise"* signal on the substrate can be coupled to the capacitor through the parasitic capacitance. Also, any voltage variation on the bottom plate of the capacitor can be coupled to the substrate and through that to other components on the chip. This communication between the capacitor

---

* *Noise* in this context refers to any undesired variation in the substrate potential, which can come from the power supply line connected to the substrate (Fig. 5.1c) or from other devices on the chip (Sec. 6.5). It should not be confused with the random noise discussed in Sec. 4.7.

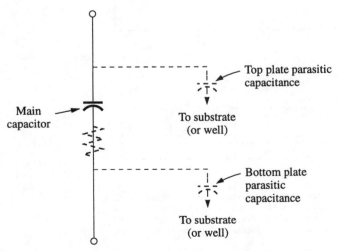

Figure 5.4 An integrated capacitor and its associated parasitics.

and the substrate can be a very serious problem (Sec. 6.5). The capacitor can be shielded from the substrate by an n well under it, which can be connected to a "quiet" dc potential [8]. This is shown in Fig. 5.3c, and it is an important technique for reducing the effects of substrate noise.

Figure 5.3d shows a different capacitor structure. In single-poly processes, this requires extra fabrication steps. It is, however, a convenient structure in some processes. Similar comments hold for the structure of Fig. 5.3e; this structure has the advantage over that in Fig. 5.3d that the n well can be used to shield the bottom plate from the substrate. For both structures, the equivalent circuit in Fig. 5.4 holds. The bottom plate parasitic capacitance is heavily voltage-dependent (being a junction capacitance) and is typically larger than the corresponding parasitic capacitance of the poly-poly or metal-poly structure (Fig. 5.3b and c) (e.g., by a factor of 2). Notice, in addition, that the bottom plate has a junction leakage current associated with it.

Variations of the structures we have presented are sometimes used, depending on the idiosyncracies of the technology employed. Some technologies, for example, allow direct contact of the top plate on top of the thin oxide, as shown in Fig. 5.5; this allows a more economical layout (smaller total layout area for a given capacitance). Other variations involve the shapes of the plates. For example, in some technologies 90° corners are avoided, as shown in Fig. 5.6. This avoids problems with imperfect etching of sharp corners, which could degrade matching between two capacitors intended to be identical. However, in some settings significant round-off errors can occur in the process from layout to mask generation, and the effect of these can be particularly bad if such 135° corners are used; in such cases, these corners will make

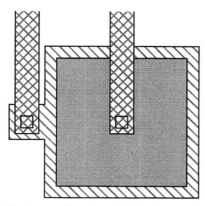

Figure 5.5 Top view of capacitor made in a technology that allows contacting over the top plate.

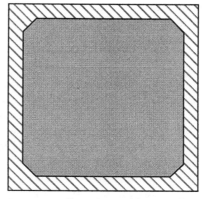

Figure 5.6 Capacitor with top plate shape without 90° corners.

matching worse rather than improve it. In general, one cannot expect that cures of certain problems in one technology can be transferred to a different one. Thus, layout guidelines and "tricks" are usually tied to a particular technology at hand and evolve after extensive experience with that technology.

In the capacitors discussed above, one or both of the plates are made of heavily doped semiconductor material. The charge in such plates distributes itself in a narrow region next to the insulator interface, in a way that is influenced by the electric field there; this, in turn, depends on the voltage across the capacitor and results in a slight dependence of the capacitance on that voltage. The heavier the doping, the more the semiconductor will behave as a metal in this respect and the smaller this effect will be. The dependence of the capacitance on the voltage across the capacitor is expressed in terms of a *voltage coefficient*, the value of which is usually given in parts per million per volt (ppm/V). A typical value for the voltage coefficient of these capacitors can be -100 ppm/V, meaning that for a 1-V change the capacitance value will decrease by 0.01 percent. Thus such capacitors can exhibit very high linearity and are widely used in data conversion and signal processing.

Similarly, the dependence of capacitance on temperature is expressed in terms of a *temperature coefficient*. For the above capacitors, a typical value for this quantity can be 25 ppm/°C.

Capacitance absolute accuracy for the above structures can be typically ± 10 percent, mainly due to oxide thickness variations; however, *ratios* of capacitances can be very accurate (for example, 0.1 to 1 percent), if the capacitors are carefully laid out in close proximity (Sec. 6.4). Also, the voltage and temperature coefficients of capacitance *ratios* will be smaller than the typical values given above for individual capacitances. Due to the popularity of switched-capacitor circuits, integrated capacitors have been the subject of extensive work [8, 10–12], especially in regard to their matching properties (Sec. 6.4).

CMOS processes developed for digital applications offer no special capacitor structure at all. Nevertheless, the need sometimes arises to design analog circuits with such processes. Junction capacitors, such as those between the p^+ layer and the n well, or the n well and the substrate, are heavily voltage-dependent and leaky due to the reverse-bias leakage current. One can use metal-poly capacitors, but the oxide between these two layers is thick, since no special fabrication step is used to provide a special capacitor oxide. This will lead to large capacitor area for a given capacitance. If this is not acceptable, one is led to using the oxide layer normally used for the gates of MOSFETs. Four possibilities are shown schematically in Fig. 5.7 [13]. The permanent connection of the substrate to $-V_{SS}$ (already encountered in Fig. 5.1c) is shown as a reminder. The structures in Fig. 5.7a and b are really MOSFETs, biased in strong inversion. Thus the bottom plate in Fig. 5.7a is formed by holes, whereas that in Fig. 5.7b is formed by electrons. Both structures are "floating" capacitors, with the one in Fig. 5.7a offering the advantage that the well can be used as a bottom shield to combat interference. The structures in Fig. 5.7c and d operate in accumulation. In Fig. 5.7c, which is a floating capacitor, $V_{GB} - V_T$ must be sufficiently *positive* to make *electrons* accumulate and form the bottom plate. In Fig. 5.7d, $V_{GB} - V_T$ must be sufficiently *negative* to make *holes* accumulate to form the bottom plate. Since the substrate is already connected to the most negative potential, biasing this structure adequately may be difficult or impossible. The permanent connection of the bottom plate to $-V_{SS}$ is a drawback anyway. In all cases, the operation of these capacitor structures is tricky since appropriate dc bias must be supplied in addition to the signal, and one must consider carefully the significant nonlinearities, parasitic capacitances, and bottom series resistance. These issues are discussed in Ref. 13.

5.2.3 Resistors

A top view of a resistor can look like that in Fig. 5.8a. A resistor with a narrow main body is used as an example. The enlarged "club heads" at the two ends are needed to accommodate the contact windows, which

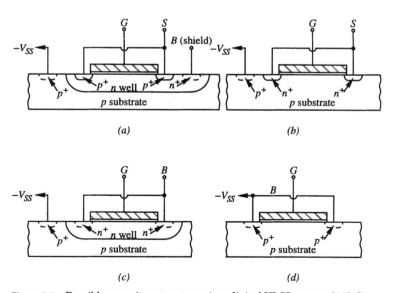

Figure 5.7 Possible capacitor structures in a digital VLSI process [13]. See text for explanations.

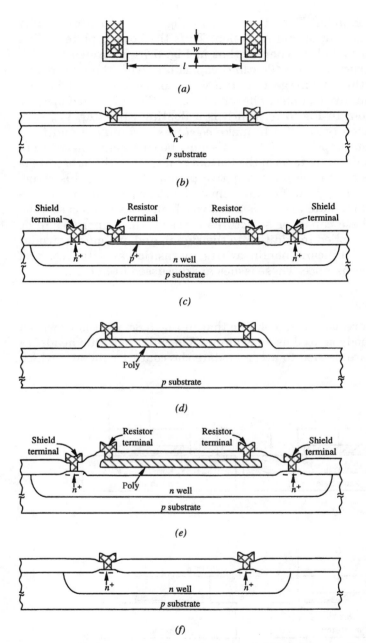

Figure 5.8 (a) Top view; and cross sections of resistors in n-well CMOS technology: (b) n^+ resistor; (c) p^+ resistor with a well as bottom shield; (d) poly resistor; (e) poly resistor with a well as bottom shield; (f) n-well resistor.

have to be a certain size (Sec. 6.3). The resistance of the *main body* is given by

$$R = R_\square \frac{l}{w} \tag{5.2}$$

with l the length and w the width, as shown, and $R_\square$ the *sheet resistance,* which is characteristic of the layer out of which the resistor is made (see below).*

When $l = w$, R is equal to $R_\square$; for this reason, $R_\square$ is often referred to as the *resistance per square* and is given in ohms per square ($\Omega/\square$). A resistor with $l = 3w$ is said to contain "3 squares" in the path of the current. To the value given by Eq. (5.2), one must add the resistances of the two ends and the associated contacts. These may not be negligible in comparison to the main resistance, especially if the resistor is short and $R_\square$ is low. Extralong resistors can be laid out as in Fig. 5.9.

The sheet resistance is poorly defined in IC processes. Uncertainties of ±20 percent (with respect to the "target" value during fabrication) are common. However, the *ratio* of two carefully laid out resistors in close proximity can be *much* more accurate (e.g., 0.2 percent). This will be discussed in detail in Sec. 6.4.

To avoid extra fabrication steps, in common processes resistors are made from the same layers that are used to make transistors. Several such resistors available in typical n-well CMOS processes are shown in Fig. 5.8b to f, and they will be discussed below [8].

n⁺ and p⁺ "Diffusion" resistors.   The resistor in Fig. 5.8b is made by using the n^+ layer. Because this layer is heavily doped, its sheet resis-

---

* For integrated resistors, $R_\square$ will be somewhat dependent on w, due to the nonrectangular shape of the resistor cross section, perpendicular to the flow of current. This will be the case especially if w is not much larger than the resistor depth. We will neglect this effect for simplicity.

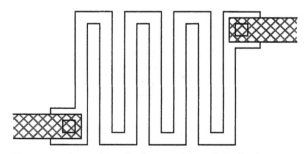

Figure 5.9   A way to make long resistors (top view).

tance is low; typical values are 30 to 100 $\Omega/\square$.* This makes the resistor unsuitable for many applications, where large resistances are needed to drop a significant amount of voltage with only limited amounts of current. With a sheet resistance of 30 $\Omega/\square$, one would need $l/w = 1000$ in Eq. (5.2), or 1000 squares, to make a 30-kΩ resistor. If the minimum width allowed is 4 µm (for reasons of matching accuracy, to be discussed in Chap. 6), a resistor 4000 µm (4 mm) long would be needed!

The reader will recall that the substrate is permanently connected to $-V_{SS}$ (Fig. 5.1c). This keeps the pn junction in Fig. 5.8b reverse-biased, which isolates the resistor from the substrate. A reverse-bias leakage current nevertheless exists. In addition, the capacitance of the depletion region is an undesirable parasitic that must be considered. A model is shown in Fig. 5.10a. The capacitance is distributed along the bottom of the resistor. The properties of such distributed structures can be described mathematically [14]. Often, they are modeled by three or more lumped elements as shown in Fig. 5.10b. The higher the frequency of operation, the larger the number of elements that must be used.

The value of the resistor depends on the bias voltage between it and the substrate, since a small part of the resulting depletion region extends into the resistor itself. This region varies in width with the applied voltage, thus varying the effective cross section through which current must flow. The heavier the doping of the region, the smaller the extent of the depletion region into the resistor (Sec. 2.4.1) and the less serious this effect will be. Although the dependence of resistance on bias is nonlinear, a rough, single value for a voltage coefficient is often quoted as an average, which assumes all parts of the resistor are at about the same voltage. The voltage coefficient for diffusion resistors can typically be 400 ppm/V. The dependence of resistance on bias results in nonlinear

---

* This value assumes that if silicides are normally used in the process (to minimize the resistance of the n^+, p^+, and poly regions), they have been prevented from forming on top of those n^+ regions that are to be used as resistors. This requires an extra mask in the process. The same is true for the p^+ and poly resistors discussed below.

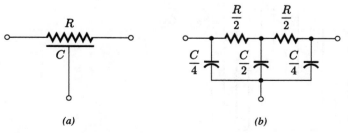

(a)   (b)

Figure 5.10 Models of an integrated resistor and associated parasitic capacitance. (a) Distributed model; (b) lumped model.

I-V characteristics; the achievable linearity should be carefully assessed before one decides to use such a resistor in a signal path.

The resistance value is also dependent (nonlinearly) on temperature; again, a single temperature coefficient is often quoted as a rough indication. A typical value of this quantity for diffusion resistors is 500 ppm/°C. The value of this coefficient is lower for heavier dopings, because then the mobility dependence on temperature becomes weaker.

Diffusion resistors exhibit *piezoresistance* effects, i.e., changes in resistance value caused by mechanical strain; such strain is present to some extent in chips after they have been packaged. The amount of this effect is different along crystallographic axes on the wafer surface.

The resistor of Fig. 5.8*b* is susceptible to interference from parasitic voltage variations on the substrate, which can come from the power supply line connected to the substrate, or from other circuits on the chip, and can cause adverse effects (Sec. 6.5). This noise can be coupled to the resistor through both the parasitic capacitance and the modulation of the resistance value by the reverse bias (as represented by the voltage coefficient). Conversely, a varying voltage on the resistor can be coupled to the substrate and through that can influence other devices on the chip. Certain other resistor structures offer a solution to these problems. One is shown in Fig. 5.8*c;* it is made of p^+ material, electrically isolated from the p substrate through an n well. The well can act as a shield, preventing noise on the substrate from being coupled to the resistor. The well is not, of course, a perfect conductor, and the shield is not perfect either; it is, however, a great help in combating interference. The n well must be connected to a sufficiently positive voltage (which must, of course, be more "quiet" than the substrate voltage) to prevent forward biasing of the junctions. A typical value for the sheet resistance of p^+ diffusion resistors is 80 $\Omega/\square$. Voltage and temperature coefficients are similar to those of n^+ resistors.

Poly resistors.   A polysilicon resistor is shown in Fig. 5.8*d*. Typical values for the sheet resistance of such structures are 25 to 100 $\Omega/\square$ (assuming no silicide layer is used on top of the resistor). The resistance value is slightly affected by the bias voltage, as the bottom of the resistor can still be depleted through the oxide field. However, the poly resistor can be placed on top of *thick* oxide. This makes its parasitic capacitance lower than that of the other resistors; the oxide field is then weak, and the voltage coefficient is low (typically 100 ppm/V). The temperature coefficient is similar to that of diffusion resistors.

The poly resistor can be shielded from the substrate by a bottom n well grounded through n^+ regions, as shown in Fig. 5.8*e*. As in the case of the resistor in Fig. 5.8*c,* this prevents undesirable communication between the resistor and the substrate.

The poly resistor is free of leakage currents. In general, this is the best resistor available in common processes, except for the fact that its sheet resistance is low and thus a large chip area is required to implement reasonable resistance values. As in the case of n^+ and p^+ resistors, an extra mask is needed to prevent silicide formation on top of those poly regions that are to be used as resistors. Sometimes special doping is used for poly resistors, resulting in higher resistivity (e.g., 1 to 2 kΩ/□); unfortunately, the temperature coefficient of such resistors is rather poor (for example, –2000 ppm/°C).

Well resistors. Figure 5.8f shows a resistor made by using the n-well layer. This layer is more lightly doped than the n^+ layer, so its sheet resistance is larger, which is in general desirable; typical values are 1 to 3 kΩ. However, its voltage and temperature coefficients are also larger (typically 10,000 ppm/V and 6000 ppm/°C, respectively). The large voltage coefficient can be reduced somewhat by making the resistor sufficiently wide that the voltage-dependent depletion regions on the sides constitute a small part of the resistor width; however, the effect of the bottom depletion region will remain. The sheet resistance can be increased somewhat by using a cap made of the p^+ layer, which is connected to a sufficiently negative voltage to prevent forward-biasing the associated pn junction. This cap reduces the cross section available to the current flow. The cap also shields the resistor from above, thus eliminating the capacitive coupling between the resistor and any interconnection lines that might be passing above it. Alternatively, the well resistor can be shielded from above by a poly layer over thick oxide (which results in low parasitic capacitance).

Other resistors. MOS transistors can be used as resistors and can be made voltage-controlled through their gate bias. Such resistors exhibit nonlinearities, and in most cases it is necessary to compensate for them. Several compensation techniques exist [15]. Also, special technologies can make available other types of resistors. For example, it is possible to deposit *thin-film* resistors on top of the oxide, by using extra fabrication steps. Although the resulting resistors have good properties, the extra cost involved can be justified only in certain cases.

5.2.4 Inductors

Integrated inductors can be made as shown in the top view of Fig. 5.11a, by using a metal layer. The inside terminal can be accessed by using a lower metal layer (broken lines), separated from the main part of the inductor by thick oxide. Several vias can be used to keep the connection resistance between the two layers low. Inductors fabricated in this way are far from ideal; a simple, approximate equivalent circuit is shown in

Fig. 5.11b. The series resistance R is the ohmic resistance of the winding and of the connections to it. The parasitic capacitances include the interturn capacitance (C_t) and the capacitance of the turns to the substrate (C_{b1}, C_{b2}). The resistors R_{b1}, R_{b2}, and R_{b3} model the substrate resistance. More accurate models use more circuit elements.

Ignoring for the moment the parasitic capacitances, we consider the *quality factor Q* of the inductor, which is defined [9] as the ratio of the

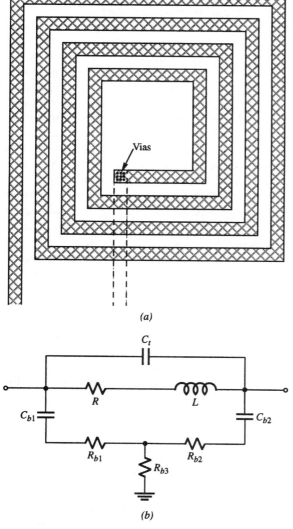

(a)

(b)

Figure 5.11   Integrated inductor. (*a*) Top view; (*b*) simple lumped model.

inductive reactance to the series resistance, or $Q = 2\pi fL/R$, where L and R are the inductance and resistance, respectively, and f is the frequency of operation. Ideally, of course, R should be zero, and then Q would be infinite. As we increase the number of turns in the inductor (keeping the metal line width fixed), both L and R increase; however, L increases faster, and thus for a given frequency Q goes up. At the same time, though, the chip area and the parasitic capacitances also increase, and a point is reached where these capacitances resonate [9] with the inductance. In the vicinity of the resonance frequency, the "inductor" does not, of course, appear as an inductor at all.

In designing an inductor, then, one must find a compromise among inductance value, quality factor, self-resonance frequency, and chip area. Although approximate analytical formulas do exist [16], an accurate evaluation requires correct inclusion of fringing field effects, the skin effect, losses due to the presence of the substrate in close proximity, etc. Extensive experimentation with test structures is common. The degrees of freedom in inductor design involve the number of turns, the width of the metal layer, the distance between turns, the outer dimensions of the inductor, and the choice of metal layer (the thicker the oxide between the inductor and the substrate, the lower the parasitic capacitance). In standard processes, the resulting inductors leave much to be desired. For example, for operation around 1 GHz one may be able to attain in a fraction of a square millimeter an inductance on the order of 10 nH with a quality factor of 3 to 5 and a self-resonance frequency of several gigahertz [17]. Typically, such an inductor would have a series resistance of 20 to 30 Ω, parasitic capacitance to the substrate of a couple hundred femtofarads, interturn capacitance of 50 fF, and substrate resistances in the hundreds of ohms. Higher Q values (for example, 10) may require chip areas as large as 1 mm^2. Better performance makes necessary the use of special techniques, such as special metallization, high-resistivity substrates, or substrate etching [18, 19].

Despite their poor quality, integrated inductors on standard silicon processes are finding some use in radio-frequency (RF) circuits for wireless communications.

5.2.5   Bipolar transistors

The layers already available in CMOS technology for forming MOS transistors can be used to form bipolar transistors as well, with no additional fabrication steps [20–23]. For n-well processes this is shown in Fig. 5.12. Two possible bipolar transistors are shown. Both are pnp transistors and use the n well as the base. If the base-emitter junction is forward-biased, bipolar transistor action is initiated.

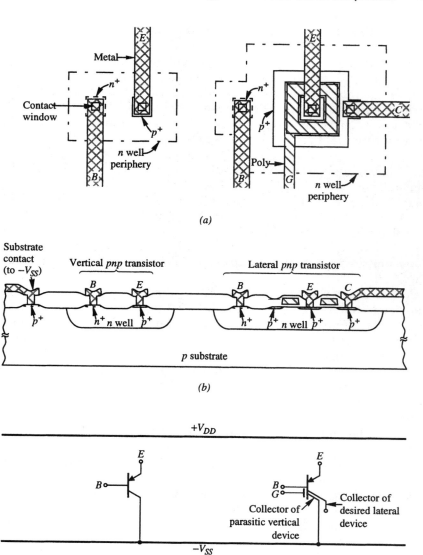

Figure 5.12 "By-product" bipolar transistors in an n-well CMOS process. (a) Top view; (b) cross section; (c) power supply connection and available device terminals.

The bipolar transistor shown on the left in Fig. 5.12b is called a *vertical* transistor, since the current in it flows vertically. The collector of this transistor is the p substrate which, as we explained in Sec. 5.2.1, is permanently connected to the negative power supply terminal $-V_{SS}$. Thus the collector of the vertical bipolar transistor is *not* available as a third terminal for circuit design purposes, as indicated in Fig. 5.12c. This limits the use of vertical bipolar devices to applications such as emitter-followers and bandgap voltage reference sources.

The transistor shown on the right [21–23] is referred to as a *lateral* transistor, since the current in it flows laterally. To ensure that this device operates as a bipolar transistor without parasitic MOS transistor action interfering with it, the gate can be biased at a sufficiently positive potential to repel the holes and to eliminate the possibility of an inversion layer at the surface [21]; the base current then flows laterally below the surface. Part of the emitter current is taken by the unavoidable vertical device formed below the emitter. To minimize this current, the horizontal emitter area should be made as small as possible. To collect as much of the current as possible, the collector of the lateral device should be laid out as a ring surrounding the emitter (Fig. 5.12a). Note that in contrast to the vertical device, the lateral transistor does make its collector available as an individual terminal (Fig. 5.12c).

Conventional CMOS processes are optimized to yield good MOS transistor characteristics, with little or no regard for bipolar transistors; the latter are really "by-products" of the CMOS process, and their performance is sometimes poor. For example, their speed can be quite low, and the range over which exponential I_C-V_{BE} behavior is observed can be quite limited. This is not always the case, though; some processes offer reasonable bipolar device performance, thus adding flexibility to circuit design. Current gain (β) values of about 100 are achievable in certain cases, with intrinsic cutoff frequencies (defined in a way similar to that in Sec. 4.5) of about 10 to 100 MHz (n-well process). In many design environments, by-product bipolar devices are not adequately characterized and are avoided. In other settings, however, such devices are used to a considerable extent.

Very good bipolar transistors can be made by using extra fabrication steps; this results in the *BiCMOS* processes, which will be discussed separately in Sec. 5.3.

5.2.6   Interconnects

In Fig. 5.1 we see that several low-resistance layers are used to make transistors. These can also be used for interconnections, and in this capacity they are often referred to as *interconnects*. The best conductor is, of course, the metal layer (its sheet resistance, e.g., can be 0.04 $\Omega/\square$); it is thus used extensively for wiring. Poly and diffusion wiring can also be used where their parasitic resistance and capacitance does not slow down circuit operation. Diffusion interconnects especially present a significant parasitic capacitance to the substrate; also, when used as part of a power supply rail, they can contribute to a serious problem known as *latch-up* (Sec. 5.2.8). Both diffusion and poly interconnects are avoided in paths where significant currents can flow (especially power supply and ground lines), due to their considerable sheet resistance. In certain analog circuits, even a few millivolts of voltage drop on wiring can destroy desirable matching between supposedly identically biased devices. In metal lines, of course, the current cannot be allowed to be arbitrarily high either; to avoid reliability problems due to a phenomenon known as *electromigration,* the current in aluminum lines is kept below a maximum allowed value per line width (typically 1 mA/μm).

Different layers must be used where two conductors must cross; an example is shown in Fig. 5.13. Depending on the fabrication process, certain crossings are not allowed. For example, the poly layer used to form self-aligned gates (Sec. 5.2.1) cannot be used to cross a "wire" made out of a diffusion layer; if such a crossing were attempted, a transistor would be created at the intersection (Fig. 5.1).

Wiring exhibits parasitic capacitance to the substrate or to other wiring next to it. Minimum-width metal or poly lines can typically have a capacitance to the substrate per unit length of 0.1 fF/μm. Two such wires made of the same layer and spaced as closely as allowed by layout rules (Sec. 6.3) can have a parasitic capacitance per unit length between them of 0.03 fF/μm. For extremely long wires, even inductive effects may have to be considered. Wiring inductance per unit length can typically be on the order of 1 nH/mm, and mutual inductance between closely spaced wires can attain a similar value; thus, long wires should be avoided in high-frequency circuits.

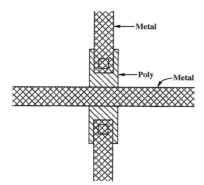

Figure 5.13 Use of a poly "crossunder" at a wire crossing.

5.2.7 Electrical parameters

The details of the fabrication process (e.g., doping concentrations, oxide thicknesses, and junction depths) determine the electrical parameters characterizing the various devices and their parasitics. An example of a set of such parameters for a typical process will now be given.

Example 5.1   A 1-μm n-well process. The minimum gate length for this process is assumed to be 1 μm, and the power supply voltage is limited to 5 V due to breakdown and device reliability considerations. Thus, this process would commonly be referred to as a 1-μm, 5-V n-well CMOS process. A "typical" value is given below for each parameter. However, the tolerances of the parameters can be wide; this will be discussed in Sec. 5.7.

MOSFET dc and low-frequency parameters at room temperature*		
	nMOS	pMOS
K'	77 μA/V^2	28 μA/V^2
V_{T0}	0.8 V	-0.9 V
γ	0.35 V$^{1/2}$	0.70 V$^{1/2}$
$\lvert \phi_0 \rvert$	0.76 V	0.72 V
$\lvert V_A \rvert / L$	21 V/μm	28 V/μm
h	2×10^{-9} V$^2 \cdot$ μm^2	8×10^{-11} V$^2 \cdot$ μm^2

Value of V_{SB}-dependent parameters at $V_{SB} = 0$:[†]

n	1.20	1.41
α	1.13	1.27
I'_M	10 nA	8 nA
k'	68 μA/V^2	22 μA/V^2

Capacitances	
Zero-biased junctions[‡]	
n^+ to substrate	
Bottom wall (per unit area)	0.31 fF/μm^2
Sidewall (per unit perimeter)	0.17 fF/μm
p^+ to well	
Bottom wall (per unit area)	0.62 fF/μm^2
Sidewall (per unit perimeter)	0.26 fF/μm
Well to substrate	
Bottom wall (per unit area)	0.28 fF/μm^2
Sidewall (per unit perimeter)	0.75 fF/μm
Gates	
Gate oxide (per unit area)	1.73 fF/μm^2
Gate poly to n^+ overlap (per unit W)	0.19 fF/μm
Gate poly to p^+ overlap (per unit W)	0.42 fF/μm
Gate-substrate overlap (per unit L)	0.30 fF/μm
Capacitors	
Capacitor oxide (per unit area)	0.8 fF/μm^2

Capacitances (*Continued*)	
Thick-oxide parasitics	
Metal-substrate (per unit area)	$0.03 \ \mathrm{fF/\mu m^2}$
Poly-substrate (per unit area)	$0.07 \ \mathrm{fF/\mu m^2}$
Metal-poly (per unit area)	$0.05 \ \mathrm{fF/\mu m^2}$

Resistances		
Layers		
n well	1000	$\Omega/\square$
$n^{+\S}$	30	$\Omega/\square$
$p^{+\S}$	80	$\Omega/\square$
Poly§	25	$\Omega/\square$
Metal	0.04	$\Omega/\square$
Contacts		
To n^+	$40 \ \Omega$	
To p^+	$90 \ \Omega$	
To poly	$30 \ \Omega$	
Vias	$3 \ \Omega$	

Leakage currents at 5-V reverse bias	
n^+ to substrate (per unit area)	$5 \ \mathrm{fA/\mu m^2}$
p^+ to well (per unit area)	$5 \ \mathrm{fA/\mu m^2}$
Well to substrate (per unit area)	$0.2 \ \mathrm{fA/\mu m^2}$

* The meaning of the symbols used here was given in Chaps. 3 and 4.

† In some cases, these values can be used even for $V_{SB} > 0$, for *rough* calculations. For more accuracy, use the formulas in Table 3.2.

‡ For a reverse-bias V_R, divide by $(1 + V_R/\phi_j)^{\eta_j}$ (Table 4.2). For n^+ and p^+ bottom wall, use $\phi_j = 0.9$ V and $\eta_j = 0.5$; for all others, use $\phi_j = 0.7$ V and $\eta_j = 0.33$.

§ The values given assume no silicide is formed; otherwise, these values reduce to $3 \ \Omega/\square$.

As illustrated by the above example, the body effect coefficient γ is larger for pMOS devices in n-well processes. The reason is that to form the n wells with a well-defined concentration, the acceptor concentration of the p substrate must be reliably overcompensated with donor atoms; the wells thus end up with a doping concentration much higher than that of the substrate. Fortunately, the body is available as a fourth terminal for pMOS devices, as we saw above; hence it can be connected to the source to make V_{SB} zero, thus avoiding the body effect, if desired. Note, though, that in such a case the well-substrate junction will be connected between source and the chip's substrate; this must be kept in mind in circuit design and simulation.

It should be emphasized again that even though one often finds the use of single "typical" values convenient, the tolerance of the parameters is actually quite poor. A discussion of this fact and typical tolerance values will be found in Sec. 5.7. The large tolerances, coupled with temperature dependencies (Secs. 3.7, 5.2.2, and 5.2.3), make it obvious that *one cannot rely on precise parameter values in integrated-circuit design.* However two identical devices located close to each other on a chip will, in general, have well-matched parameter values (Sec. 5.7); this has led to several circuit techniques for obtaining precision performance by relying only on matching.

5.2.8   Input protection

As mentioned in Sec. 3.9, "static" electricity can charge the transistor gates connected to input pads, to the point that oxide breakdown can occur. To prevent this, protection circuits are connected to such pads. These can have various forms [4, 24–27] and are often referred to as *electrostatic discharge (ESD)* protection circuits. A common approach is to place a poly resistor between the input pads and the gates to be protected, and to connect reverse-biased junctions between those gates and the power supply lines. If the reverse breakdown voltage of the junctions (which is lower than the gate oxide breakdown voltage) is exceeded, the junctions will conduct and their current will cause a voltage drop across the poly resistor, thus limiting the voltage seen by the gates. At the output of a circuit, if the pad is connected to the drains and/or sources of large devices, the *pn* junctions associated with such devices act as protection circuity.

Input pad protection circuits often differ from process to process and from manufacturer to manufacturer. Such circuits should be designed by someone with intimate knowledge of the fabrication process and in conjunction with strategies for latch-up prevention (see below). Information about protection circuits (in fact, complete input and output pad cells which include protection) is usually provided by the fabrication houses that a designer is dealing with. Some of the pads developed for digital circuits can cause problems in analog work (e.g., in terms of leakage or linearity). It is thus important to consider the effect of the pads in the design process.

5.2.9   Latch-up

A problem with CMOS processes is the *latch-up* condition. If a *pn* junction is accidentally forward-biased, the *pnpn* structures in Fig. 5.1b can be activated and can act as *silicon controlled rectifier (SCR)* devices [4, 24–29]; large currents can then flow, with *pn* junction forward-biases supported across well and substrate resistances. These currents prevent proper operation and can destroy the circuit, if they become excessive. Latch-up can be activated through sudden transients that can forward-bias a junction; these transients can occur when power supplies are turned on as a result of switching transients or ringing in the power supply and ground lines. Latch-up can also occur when various voltages inside a chip are established in a wrong, unintended order, such as when the power supplies are switched in the wrong order, or when a chip is replaced on a board, or a board on a system, while the power is on ("hot" plug-in). It can also occur as a result of "static" charge on the body of a human handling a circuit or as a result of currents induced by radiation. The strongest suspects for causing latch-up are the devices connected to input/output pads, where voltages can exceed their normal range during transients and where large currents can flow.

Latch-up is controlled by both fabrication processing and layout techniques [4, 24–29]. In certain modern n-well processes, the p substrate is a layer formed by a process called *epitaxy* [1–5] on top of a heavily doped p-type wafer. The heavy doping in the latter helps keep the parasitic substrate resistance low, thus limiting parasitic voltage drops that could forward-bias junctions and activate latch-up. Guidelines for layout, appropriate for avoiding latch-up in the process at hand, are usually supplied by fabrication houses. These can include, for example, the use of sufficient spacing between the edges of the n well and the sources or drains in it, to lower the β of the parasitic bipolar transistors; the use of multiple contacts to the wells and substrate, to lower the parasitic resistances; and the use of *guard rings* around suspect devices and wells, to lower parasitic resistance values and to collect carriers (e.g., in an n-well process one can use p^+ rings connected to $-V_{SS}$ around nMOS devices, n^+ rings connected to $+V_{DD}$ along the periphery of n wells, and n-well rings connected to $+V_{DD}$ surrounding normal n wells). Sometimes an entire area of the chip, containing many circuits, is designed with no special latch-up protection circuits in it and is heavily guarded as a whole against latch-up by using guard rings all around the area. Contacting the back of the chip can be a significant help, too, although such contact may not be possible with certain packages or during wafer-level testing.

5.3 BiCMOS Processes

5.3.1 Adding high-performance bipolar transistors to CMOS

In the CMOS process we discussed in the previous section, the bipolar transistors available were seen to be just by-products; such processes are optimized to provide good MOS transistors, and as a result the performance of the bipolar by-products is poor. However, with extra fabrication steps one can create processes that make possible both CMOS devices and *optimized* bipolar transistors. These are called *BiCMOS* (bipolar CMOS) processes. Such processes [30] add flexibility to circuit design at an increased fabrication cost. As an example, consider the *n*-well CMOS process of Fig. 5.1 as a starting point; in principle it can be

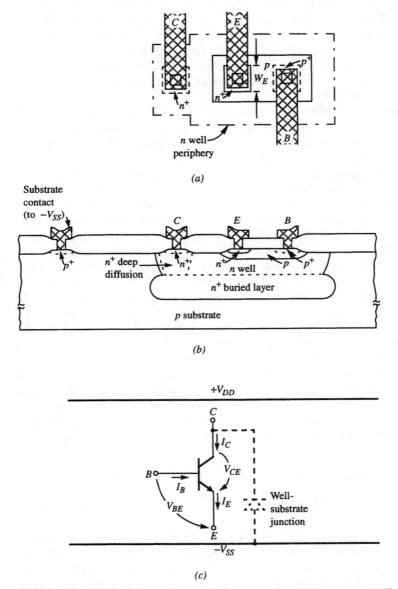

Figure 5.14 An *npn* bipolar transistor in a simple BiCMOS process. (*a*) Top view; (*b*) cross section; (*c*) device symbol and definition of voltages and currents (the diode shown models the well-substrate junction).

converted to a BiCMOS process by adding fabrication steps to produce an *npn* bipolar transistor on the same chip, as shown in Fig. 5.14*a* and *b*. An extra *p*-type diffusion forms the transistor's base. An n^+ buried layer is used to make possible a low collector series resistance. To further lower this resistance, a deep n^+ diffusion can be used, as shown. In contrast to the vertical by-product bipolar transistor in standard CMOS processes (Sec. 5.2.5), here all terminals of the device are available to the circuit designer. This is indicated in Fig. 5.14*c*. The well-substrate junction accompanies the device as shown, since the *p*-type substrate is permanently connected to $-V_{SS}$ (see also Fig. 5.1). BiCMOS processes with cutoff frequencies in the 5- to 10-GHz range are common, and 0.5-μm processes with f_T exceeding 30 GHz are in use. An example of a modern BiCMOS process is shown in Fig. 5.15 [31]; this process uses a *polysilicon emitter* for improved current gain. Many other variations of the BiCMOS process are in use [30]. Figure 5.16 shows a *double-polysilicon* bipolar structure; such structures allow self-alignment during fabrication for reduced dimensions and small parasitic resistances and capacitances [32].

Structures for *pnp* devices can be made along the lines discussed in Sec. 5.2.5. In the case of a lateral *pnp* device (right in Fig. 5.12*b*), the addition of a buried layer below the base can help reduce the parasitic vertical action. However, *pnp* devices made in this way are still by-

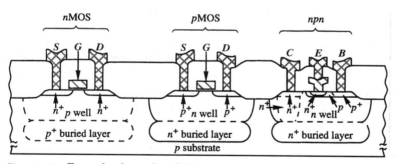

Figure 5.15   Example of a modern BiCMOS process.

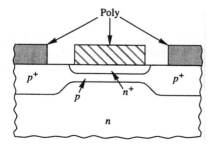

Figure 5.16   A double-polysilicon self-aligned bipolar transistor structure.

products of the process, unlike the *npn* device described above, and thus their performance is limited ($f_T < 100$ MHz), as discussed in Sec. 5.2.5. With extra fabrication steps, optimized *pnp* devices are possible ($f_T > 1$ GHz) and are sometimes used [31].

5.3.2 Bipolar transistor models

Bipolar transistors are discussed in detail in many books [33–38], and adequate summaries of bipolar transistor operation and modeling can be found in circuit design texts [39]. For this reason, our exposure to bipolar transistors will be limited to a brief review of some important facts. The voltages and currents used in the following discussion are defined in Fig. 5.14c.

Large-signal characteristics. An example of I_C-V_{CE} characteristics for an *npn* transistor with $V_{CE} \geq 0$ is shown in Fig. 5.17a (with V_{BE} as a parameter). In analog circuits, by far the most important operating region is the *forward active* region, which is basically the region in which the I_C-V_{CE} characteristics shown become flat.* This requires V_{CE} values of a few tenths of a volt or more, depending on transistor used, bias, and circuit function. For common I_C values in the forward active region, typical devices require V_{BE} values of about 0.6 to 0.85 V. Characteristics with I_B as a parameter are shown in Fig. 5.17b.

Important relations between large-signal currents and voltages in the *forward active region* are summarized in the first part of Table 5.1. The quantity I_S depends strongly on the fabrication process (a typical value is 10^{-17} A for a minimum-size device) and is strongly temperature-dependent (it doubles every 8 to 10°C increase). The quantity ϕ_t is the thermal voltage, which is proportional to the absolute temperature (see Table 3.3). The combined effect of the temperature dependence of I_S and ϕ_t is such that, for constant V_{BE}, the collector current I_C increases with increasing temperature; for constant I_C, V_{BE} decreases almost linearly with increasing temperature (typically by 2 mV/°C). The dependence of bipolar transistor characteristics on temperature is discussed in detail elsewhere [40]. The quantity β_F is the large-signal dc forward current gain; it is typically in the range of 50 to 200 and increases with temperature (typically by 0.5 to 1 percent/°C). Both I_S and β_F are poorly controlled during processing; if β_F is said to have a typical value of 100, its actual value could be, for example, anywhere between 70 and 140.

---

* In the case of the bipolar transistor, this region is *not* called "saturation." In fact, this name traditionally is reserved for the region seen to the left of the forward active region, where I_C is strongly dependent on V_{CE}.

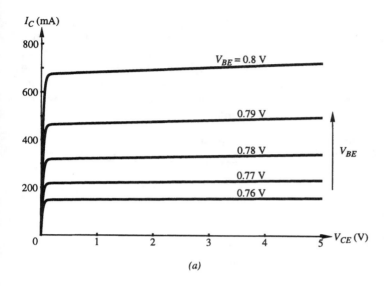

(a)

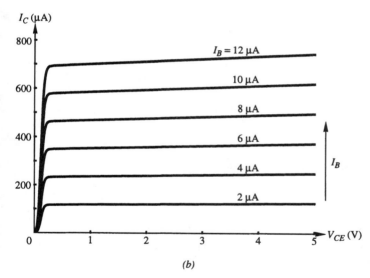

(b)

Figure 5.17 Typical *npn* bipolar transistor V_{CE}-I_C characteristics with (*a*) V_{BE} as a parameter, (*b*) I_B as a parameter.

Figure 5.18 shows typical plots of I_C and I_B (on a log scale) versus V_{BE}. Plots using the expressions of Table 5.1 are shown in broken lines for comparison (a constant β_F is assumed in the expression for I_B). The limitations on the accuracy of these expressions become apparent in the figure. The base current I_B deviates from ideality at low currents (due to recombination currents in the base-emitter space-charge layer [33–38]). The expression for I_C is valid over several decades, but at high currents this validity is lost due to several effects, including high-level injection, base stretching (Kirk effect), and lateral base widening [33–38]. The high-current effects also influence the base current. Internal voltage drops can develop across the resistance of the base material (which resistance is markedly bias-dependent).

TABLE 5.1 Important Relations for an *npn* Bipolar Transistor in the Forward Active Region

Large-signal relations
$I_C = I_S \exp(V_{BE}/\phi_t)$ *
$I_B = \dfrac{I_C}{\beta_F}$
$I_E = I_C + I_B = (\beta_F + 1)I_B$
Small-signal parameters
$g_m = \dfrac{I_C}{\phi_t}$
$g_\pi = \dfrac{g_m}{\beta_0}$
$g_o = \dfrac{I_C}{V_A}$
$g_\mu = \dfrac{g_o}{\zeta\beta}$ †
$C_\pi = C_{je} + C_b$
$C_b = \tau_F g_m$
$C_{je}, C_\mu, C_{cs}, C_{bs}$ junction capacitances‡
Noise power spectral densities§
$S_{I_c} = 2qI_C$
$S_{I_b} = 2qI_B + \dfrac{FI_B^a}{f}$ ¶

* If it is desired to include the dependence of I_C on V_{CE}, one can use $I_C = I_S \exp(V_{BE}/\phi_t)(1 + V_{CE}/V_A)$. Leakage currents and breakdown effects are not included. The expression is not valid at very high currents.

† The quantity ζ is typically between 2 and 15.

‡ For the *npn* device, $C_{bs} = 0$; for a lateral *pnp* device, $C_{cs} = 0$. The quantities C_{je}, C_μ, C_{cs}, and C_{bs} are reverse-biased junction capacitances and are of the form $C_j = C_{jo}/(1 + V_R/\phi_j)^{\eta_j}$, where C_{jo} is the zero-bias capacitance value, ϕ_j is the junction built-in potential, and η_j is a characteristic exponent. In the forward active region, the base-emitter junction is forward-biased and C_{je} is difficult to calculate by hand, so simple estimates are often used (e.g., double the zero-bias capacitance).

§ Here q is the electron charge (1.602×10^{-19} C), and F and a are process-dependent parameters.

¶ This model assumes that a type of noise known as *burst* noise [39] is negligible; if this is not the case, a contribution of the form $GI_B^b[1 + (f/f_b)^2]^{-1}$ should be added to S_{I_b}, with G, b, and f_b appropriate constants.

With β_F defined as I_C/I_B, we have $\log \beta_F = \log I_C - \log I_B$; that is, $\log \beta_F$ corresponds to the spacing of the two curves in the semilog plot of Fig. 5.18. If β_F is plotted versus I_C then, we obtain a plot of the type illus-

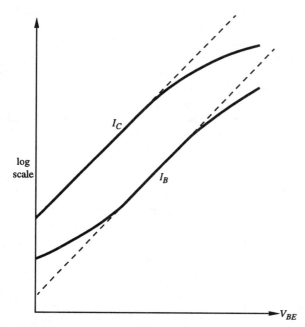

Figure 5.18 I_C and I_B (log scale) versus V_{BE}.

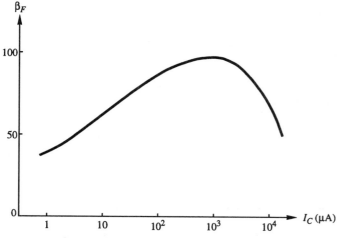

Figure 5.19 β_F versus I_C.

trated in Fig. 5.19. It is apparent that only for medium-valued currents can β_F be assumed to be relatively constant.

As mentioned in Table 5.1, the slope of I_C versus V_{CE} in saturation can be modeled by multiplying the current expression given there by $1 + V_{CE}/V_A$, where V_A is known as the *Early voltage* and is bias-dependent (although it is often assumed to be constant in the literature).

If V_{CE} is not sufficiently large, at high values of I_C the voltage drop across the internal resistance of the collector material can cause problems. This drop subtracts from V_{CE}. Thus the transistor sees internally a smaller collector-emitter voltage than the externally applied V_{CE}, and can enter saturation even at rather large values of the latter. This effect is known as *quasi-saturation* and is illustrated in Fig. 5.20. The shape of the characteristic in the quasi-saturation region is complicated by the fact that the internal resistance causing this effect is bias-dependent.

Small-signal models. A small-signal equivalent circuit for the forward active region is shown in Fig. 5.21, with expressions for the model parameters given in Table 5.1. All quantities shown next to resistors in the figure are conductances. Quantities r_b, r_c, and r_{ex} correspond to the physical resistances of the base, collector, and emitter regions, respectively. In contrast to these, g_π, g_μ, and g_o represent small-signal current-voltage relations describing the intrinsic transistor action (just as g_{ds} does for the MOS transistor—Sec. 4.2) and do not represent physical resistive regions. In the expression for g_π in Table 5.1, the quantity β_0 is the *small-signal current gain*. This quantity is close to β_F in the cur-

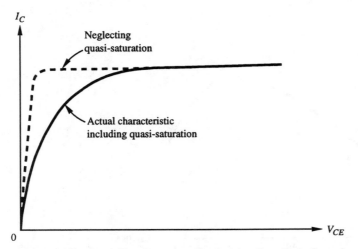

Figure 5.20 I_C versus V_{CE}, for a given I_B, showing the effect of quasi-saturation.

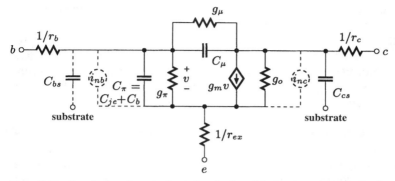

Figure 5.21 Small-signal equivalent circuits for a bipolar transistor in the forward active region. All quantities shown next to resistors are conductances.

rent region where β_F is approximately constant (center region in Fig. 5.19), but can differ from β_F significantly outside this region.

The quantity C_b is an intrinsic transistor capacitance called the *diffusion capacitance*. In the expression for this parameter in the table, τ_F is the *forward transit time*. The quantities C_{je}, C_μ, C_{cs}, and C_{bs} are the reverse-bias capacitances of the junctions formed between emitter and base, base and collector, collector and substrate, and base and substrate, respectively.* For the *n*-well BiCMOS process we have discussed, C_{cs} will be present in the device of Fig. 5.14*b*, whereas C_{bs} will be zero. The opposite will be true for devices where the base forms a junction with the substrate, such as the *lateral pnp* transistor in Fig. 5.12. Junction capacitances can be calculated by considering bottom-wall areas and sidewall perimeters, as was done for the MOS transistor in Sec. 4.4.1.

For noise calculations (Sec. 4.7.1), two intrinsic noise current sources must be included in the model, as shown in Fig. 5.21: i_{nc}, which represents the shot noise in the collector current, and i_{nb}, which includes the shot noise and $1/f$ noise in the base current. The power spectral densities for these noise currents are given in Table 5.1. In addition, noise is, of course, contributed by the physical resistors r_b, r_c, and r_e (but *not* by the g_π, g_μ, and g_o elements which do not correspond to physical resistive regions); the corresponding noise sources can be modeled as explained in Sec. 4.7.1, but are not included in the figure to avoid making it crowded.

In many cases, adequate small-signal calculations can be performed using a simplified model, resulting from the one in Fig. 5.21 after removal of certain elements (usually g_μ, r_{ex}, and r_c).

---

* The subscripts b and s here stand for *base* and *substrate*, respectively, and should not be confused with the same subscripts used in Chaps. 2 to 4 for the body and source of a MOS transistor.

A note of caution concerning g_o is in order. If a small value is desired for this quantity (which is essential for high-gain amplification), the bias and signal combinations used should be such that the transistor operates always in the flat part of the I_C-V_{BE} characteristics in Fig. 5.17. Depending on device construction and collector current, this may require V_{CE} values significantly larger than the few tenths of one volt commonly assumed (e.g., V_{CE} may have to be as large as V_{BE}).

Behavior at very high frequencies.   As was the case for the MOS transistor (Sec. 4.6.2), at very high frequencies the bipolar transistor exhibits nonquasi-static effects and the model of Fig. 5.21 becomes inadequate. To take into account such effects [34], one should replace the transcon-

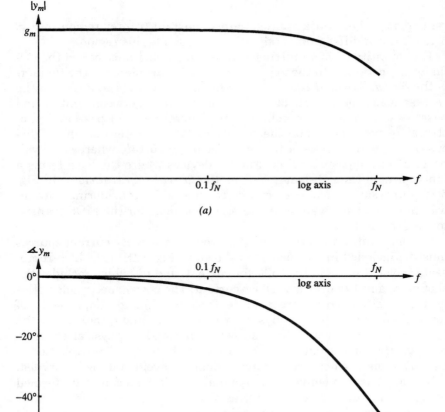

Figure 5.22 Bipolar transistor transadmittance (a) magnitude and (b) phase.

ductance source $g_m v$ in Fig. 5.21 by a *transadmittance* source $y_m V$, where V is the phasor [9] corresponding to v and y_m is the transadmittance, given by

$$y_m = \frac{g_m}{1 + j(f/f_N)} \qquad (5.3)$$

with $f_N \simeq 3(2\pi\tau_F)^{-1}$, where τ_F is the forward transit time. The behavior of y_m (magnitude and phase) versus frequency is shown in Fig. 5.22. As will be seen shortly, f_N is at least 3 times larger than the unity-gain frequency of the device.

The small-signal current gain of the bipolar transistor can be defined with the help of Fig. 5.23, where I_B and I_C are bias currents. In the sinusoidal steady state [9], let I_b and I_c be the complex phasors representing $i_b(t)$ and $i_c(t)$, respectively. We define the small-signal current gain as the ratio I_c/I_b, and we denote it by $\beta(f)$. In general, this quantity is complex, and it behaves as shown in Fig. 5.24. Here the broken lines are obtained from the model in Fig. 5.21 (neglecting g_μ, r_{ex}, C_{bs}, and r_c), whereas the solid lines are obtained if in that model $g_m V$ is replaced by $y_m V$, with y_m as given by Eq. (5.3) (Prob. 5.4). As seen, the model of Fig. 5.21 cannot be trusted at very high frequencies, due to the presence of nonquasi-static effects. We can nevertheless define a *unity-gain frequency* or *transition frequency* f_T as shown, as the frequency at which $|\beta(f)|$, as predicted by our simple model, drops to unity. The value of f_T obtained in this way is (Prob. 5.4)

$$f_T \simeq \frac{g_m}{2\pi(C_\pi + C_\mu)} \qquad (5.4)$$

which, by using $C_\pi = C_{je} + C_b$, $C_b = \tau_F g_m$, and $g_m = I_C/\phi_t$ from Table 5.1, can be written as a function of I_C as follows:

$$f_T = \frac{1}{2\pi[\tau_F + (\phi_t/I_C)(C_{je} + C_\mu)]} \qquad (5.5)$$

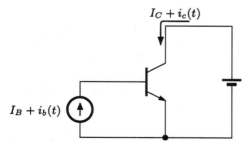

$I_C + i_c(t)$

$I_B + i_b(t)$

Figure 5.23 Circuit used in the definition of the small-signal current gain.

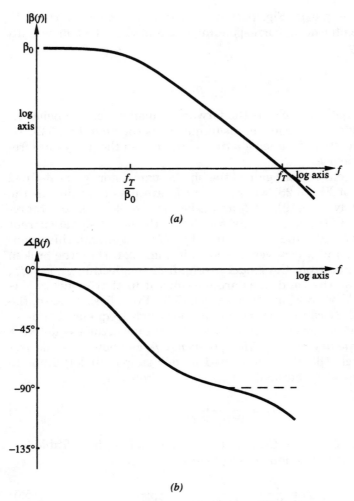

Figure 5.24 Small-signal current gain of common-emitter connection versus frequency (log axes). (*a*) Magnitude; (*b*) phase.

Thus, as I_C is increased, starting from low values, f_T rises and tends toward a maximum value of $f_{T0} = 1/(2\pi\tau_F) \approx f_N/3$. This is shown in Fig. 5.25. Increasing V_{CB} helps raise f_T, by decreasing C_μ in Eq. (5.5). As shown in the same figure, at higher I_C values f_T drops again because τ_F begins to increase, due to high-current effects. These effects have been seen above to contribute to a decrease in the value of β_F at high currents. The operating point of bipolar transistors is usually chosen such that both reasonable β values and high speed can be obtained. This point corresponds to a given emitter current per unit of emitter area (say, 200 μA/μm^2).

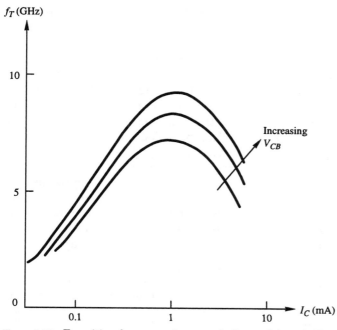

Figure 5.25   Transition frequency f_T versus I_C (log scale), with V_{CB} as a parameter.

It can be seen from Figs. 5.22 and 5.24 that, even at frequencies as high as f_T, the magnitudes of y_m and of $\beta(f)$ are predicted satisfactorily by the model in Fig. 5.21 (recall that f_T can be at most equal to $f_{T0} = 1/2\pi\tau_F$). The phase of these quantities, though, is not predicted accurately at such frequencies. The error at f_T is maximum when the device operates at optimum current, since then f_T can approach $f_{T0} \approx f_N/3$; it can be calculated that in that case, the excess phase caused by nonquasi-static effects at $f = f_T$ is about $-18°$. Such errors can be serious in some contexts, e.g., feedback amplifier design. The modeling at such high frequencies is further complicated by the distributed nature of the various resistances and capacitances. Models used in computer simulators are sometimes augmented by external resistors and capacitors, chosen so that they can approximately model the various effects discussed above. The excess phase is sometimes represented by a complex factor multiplying $\beta(f)$.

The unity current gain frequency f_T is not an adequate figure of merit for high-frequency performance. Perhaps more important is the unity power gain frequency f_m, sometimes called *maximum oscillation frequency*, which can be shown to be given by [33]

$$f_m = \sqrt{\frac{f_{T0}}{8\pi r_b C_\mu}} \tag{5.6}$$

where f_{T0} is again the maximum (theoretically) achievable f_T. Note that transistors designed for maximum f_T are not necessarily optimum in terms of f_m.

If larger currents need to be handled, the emitter area can be increased to maintain the current density near its optimum value. The simplest way to do this is to increase W_E in Fig. 5.26. The horizontal dimensions are kept at their minimum value, so all resistances are kept small; in fact, it is easy to see that the resistances are approximately *inversely* proportional to W_E (Prob. 5.5), provided the contact resistance is reduced accordingly by contacting the n^+ and p^+ strips over most of their length (a way to do this will be discussed in Sec. 6.4). Also, if W_E is rather large, it is easy to see that all capacitances are approximately *proportional* to W_E. Thus all *RC* products are approximately independent of W_E and so is, to a first order, the frequency performance of the device, provided the optimum current density is maintained. Other ways to lay out bipolar transistors will be considered in Chap. 6.

Example 5.2 A 1-μm 5-V BiCMOS process. We now give an example of a 1-μm BiCMOS process. For simplicity, the parameters of the *n*MOS and *p*MOS devices, layer resistances, parasitic capacitances, etc., in this process will be assumed to be the same as for the CMOS process of Example 5.1. The parameters of an *npn* bipolar transistor with a 9-μm² emitter area, operating at room temperature, are given:

β_F	85
I_S	3×10^{-18} A
V_A	75 V
τ_F	15 ps
r_b	140 Ω

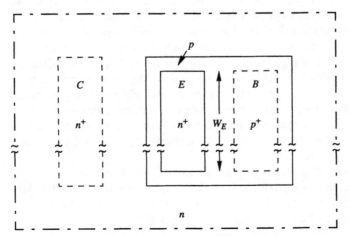

Figure 5.26 A way to increase the emitter area of bipolar transistors by scaling the emitter width W_E.

r_c	18 Ω
r_{ex}	9 Ω
C_{je}	24 fF *
C_μ	18 fF †
C_{cs}	54 fF †
F	3×10^{-15} A
a	1*

* The value given is an average for typical values of V_{BE} in the forward active region. Since in this region the base-emitter junction is forward-biased, its capacitance as a function of V_{BE} is difficult to calculate by hand. The above value can be used as an estimate.

† The values given are for zero junction bias. For C_μ or C_{cs} under reverse bias, divide by $(1 + V_R/\phi_j)^{\eta_j}$, where V_R is the reverse-biased voltage; for C_μ, $\phi_j = 0.55$ V and $\eta_j = 0.3$; and for C_{cs}, $\phi_j = 0.6$ V and $\eta_j = 0.4$.

Models for *pnp* transistors are easily produced by adopting the above results, in a way similar to that suggested in Sec. 3.10 for *p*MOS transistors. In the forward active region, V_{BE}, V_{CE}, I_B, I_C, and I_E are all negative. The small-signal equivalent circuit of Fig. 5.21 remains valid, except that for a lateral *pnp* device C_{cs} is zero, whereas C_{bs} is nonzero (see Fig. 5.12). All parameters in this circuit are positive. Small-signal parameter values can be obtained from the expressions in Table 5.1, if in them one replaces dc bias currents and voltages by their absolute values.

5.4 Other Silicon Processes

***p*-Well CMOS processes.** The *n*-well CMOS process we have discussed above has its complement; this is shown in Fig. 5.27 and is referred to as a *p-well process* [1–6]. The substrate here is *n* type, and the individual wells are *p* type. This results in *p*MOS devices on a common substrate (which is tied to $+V_{DD}$ to prevent forward-biasing of *pn* junctions), and *n*MOS devices in *p* wells (one or more devices in each well). The *n*MOS devices end up having a larger body-effect coefficient, due to their higher body doping. The *p*-well process also provides, as byproducts, *npn* bipolar transistors. The latter can typically have current gains of 100 and unity-gain frequencies on the order of 30 to 300 MHz.

While both *n*-well and *p*-well processes have been used widely in the past, today *n*-well processes are dominant. In analog work, *n*-well processes are attractive because the *p*MOS devices (which can be advantageously used as input devices in amplifiers, due to their low noise) are inside wells, and thus their body does not have to be connected to a (usually noisy) power supply line; it can instead be tied to the transistor sources. On the other hand, their junction parasitic capacitances are larger than those of *n*MOS devices, because the *n* well is doped more heavily than the substrate. Given that *p*MOS devices are intrinsically "slower" anyway, due to their lower mobility, the additional slowdown due to the large capacitances is certainly undesirable.

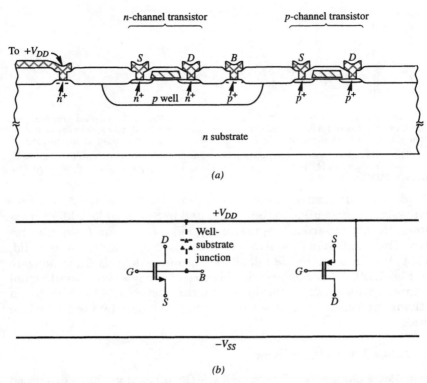

Figure 5.27   MOS transistors in a p-well CMOS process. (a) Cross section; (b) power supply connection and available terminals.

Twin-well CMOS processes.   A CMOS process called *twin-well, dual-well,* or *twin-tub* provides individually optimized bodies for both nMOS and pMOS devices, on top of lightly doped *epitaxial* material formed on a heavily doped wafer [1–6]. In this way, the characteristics of each type of device can be better controlled individually; e.g., if desired, the two types can be made to have about the same body-effect coefficient. It should be emphasized, though, that the presence of individual wells does not mean that the body of both nMOS and pMOS devices is available as a fourth terminal. For example, in a p-substrate twin-well process, the nMOS devices are sitting inside p wells, which are thus electrically connected to the p substrate. Thus, in this example the availability of device terminals is the same as in Fig. 5.1c, and circuit design considerations are similar to those in the case of simple n-well processes.

Other MOS processes.   Special CMOS technologies are used to minimize device-to-device distance (*trench isolation technology*), improve device isolation and eliminate latch-up [*silicon-on-insulator* (SOI)], etc. [1–5]. Such technologies will not be considered here. Also, older single-channel technologies will not be considered; these include the NMOS technology, in which usually both enhancement- and depletion-type nMOS devices are available, and the veteran PMOS technology. These technologies, especially NMOS, were dominant in the first several years of analog MOS LSI. Today CMOS and BiCMOS are well established as the most appropriate VLSI technologies, so they are the only technologies discussed in this book.

Bipolar processes. *Bipolar processes* are those that make available only bipolar transistors (no MOS devices) [1–3]. Such processes have been in use for a long time, and today are used mainly for continuous-time analog circuits and for high-speed digital circuits. The scale of integration of such circuits, though, is often limited due to large chip area and power dissipation per logic gate. This has limited the activity in the area of mixed analog-digital bipolar VLSI circuits. Bipolar transistors do have attractive properties for certain kinds of analog circuits. With the advent of BiCMOS processes, though, these properties can be taken advantage of without having to switch to fully bipolar technology, and thus the best of both worlds can be combined.

Other possibilities. Today fabrication technology has reached a mature stage in which very fancy processes are possible. With extra fabrication steps, other device combinations, in addition to those mentioned so far, can be had. Thus, e.g., it is possible to make two kinds of nMOS transistors on the same chip, one with low and one with high threshold, for added circuit design flexibility. There have even been reported processes that make six types of devices available on the same chip: nMOS enhancement, nMOS depletion, pMOS enhancement, pMOS depletion, bipolar *npn,* and bipolar *pnp.* Such processes are more expensive than the ones we have discussed so far and are not used extensively. It is also possible to modify existing processes to add certain features, such as thin-film resistor capability and lightly doped poly. Any such modification adds to the cost of the process.

Special processes are used for high-voltage and power circuits. Such processes will not be covered here; the interested reader can consult Refs. 41 and 42.

5.5 Sensors

Integrated circuits are often employed in applications where the signals to be processed are not electrical at their sources of origin. They can be, for example, temperature, light intensity, magnetic field, mechanical pressure, etc. Sensors are used for converting such signals to electrical form, so that they can be subsequently processed by electronic circuits. A variety of technologies for making sensors have been developed, but many are not compatible with VLSI technology. However, some sensors can be made on VLSI chips [43–48]. For example, the thermal properties of on-chip devices, notably bipolar transistors, have been used to make thermal sensors for many years. Optical sensors can be made by using the optical properties of *pn* junctions and transistors to sense light intensity. The Hall effect on transistors can be used to make magnetic field sensors. Mechanical sensors can also be made, by using device mechanical properties (such as piezoresistance) to sense strain, pressure, acceleration, etc.

Since VLSI fabrication technology is normally not optimized for making sensors, on-chip sensors are by-products of the process and their performance is not always satisfactory. Fortunately, this can be often compensated for by using sophisticated electronic means, leading to on-chip *smart sensors.*

5.6 Trimming

For high-precision applications, *trimming* is sometimes used; this refers to the adjustment of element values on the chip after fabrication, by a "surgical" procedure. For example, laser beams are sometimes used to modify thin-film resistor values or to break or make connections. Or, current pulses are used to blow *polysilicon fuses;* this can be used to either disconnect parts of resistor or capacitor groups or to activate or deactivate MOS transistor switches, which in turn connect or disconnect parts of resistor or capacitor groups (it must, of course, be ensured that the switch resistance does not interfere with proper operation). In a different trimming process, currents can be used to cause a permanent short circuit in base-emitter zener diodes (*zener zapping*), thus short-circuiting parts of a resistor or capacitor array.

An alternative to such trimming is to use a process that allows the fabrication of electrically erasable programmable read-only memory (EEPROM) devices on the same chip [49]. Such devices use floating gates to store charges that are placed on them through high-voltage pulses, which in a sense cause a temporary breakdown for this purpose. The high-voltage pulses can be created on the chip from the normal power supply voltage through voltage multiplication. EEPROM devices can be used to store logic levels which are in turn used to close or open analog switches; thus, elements can be added to, or removed from, element arrays. The arrays can consist of capacitors, resistors, or transistors. Since the programming of the EEPROM devices is done electrically, the trimming can be done on the final, packaged devices. The advantage is that errors that can occur due to packaging (e.g., due to mechanical strain) can be corrected, and the trimming can be updated as needed to correct for changes in temperature, humidity, aging, etc. EEPROM devices which can be reprogrammed millions of times are possible, with data retention times projected to be longer than 20 years. In addition to storing digital levels, such devices have been used to store analog levels (i.e., values from a continuum between zero charge and maximum charge) on their gates. In this way, for example, the channel resistance of MOS transistors can be programmed over a wide range [50–52].

Both fabrication "fanciness" and postfabrication surgical trimming can add significantly to the final cost of a chip. They are usually avoided unless it is clear that the desired result cannot be achieved by simpler means. A lot of circuit design ingenuity has gone into making possible standard-process, untrimmed circuits with high performance. Key to making this possible is an understanding of the accuracy and matching of electrical parameters in integrated circuits. The next section deals with this topic.

5.7   Tolerance and Matching of Electrical Parameters

A number of uncertainties during fabrication (e.g., as to the exact furnace temperature or the ion implantation dose) have as a result significant tolerances in the process parameters, for example, ±5 percent for the oxide thickness or ±10 percent in the body doping. These, in turn, imply corresponding tolerances in the electrical parameters characterizing the various devices.* For example, for a typical process, tolerances for the threshold voltage might be ±0.1 V; for the parameter K', ±10 percent; for the body effect coefficient γ, ±10 percent; for capacitances per unit area, ±5 percent; for resistances, ±20 percent.† Geomet-

---

* Often parameter values are given in the form $\alpha \pm \Delta\alpha$, where the tolerance $\Delta\alpha$ corresponds to a specified number of standard deviations, e.g., three. Sometimes, though, a parameter's statistical distribution cannot be assumed to be symmetric around a mean, so plus-or-minus tolerance intervals cannot be used. This is especially so for some parameters with large variations; e.g., in bipolar transistors β may have a typical value of 100, with probable values ranging from 50 to 200.

† The numbers given are only intended as a rough indication. The tolerances will be different for different types of transistors or resistors in the same process, and of course for different processes. Also, a designer must at times work with much wider tolerance intervals than the above (e.g., by a factor of 2 or more) for the following reason. In certain design environments, a design is submitted to an entity acting as an interface between designers and chip makers (silicon foundries). For reasons of scheduling, pricing, availability, etc., the design may be submitted by this "interface house" to any of a number of different chip makers, which use similar—but not identical—processes. Since it is usually not known ahead of time which chip maker will fabricate the design, the tolerances that the designer must assume are a combination of the tolerances of all potential manufacturers involved, and thus they are large.

ric layout dimensions will also exhibit tolerances; for example, channel lengths and widths may have tolerances of ±0.2 μm.

Despite the above facts, designers often use—among other pieces of information—sets of "typical" electrical parameters, like those in Examples 5.1 and 5.2. These do, indeed, provide some rough estimate of what can be expected, and in so doing they serve a useful purpose. At the same time, it should be realized that only if all fabrication parameters involved happen to be at their mean value can one expect to find devices characterized by typical electrical parameters; and this is extremely unlikely. *Relying on precise electrical parameter values for proper integrated circuit operation is absurd.* Thus, tolerances must be taken into account in professional design.

The problem with using independent tolerance values, such as those given above, is that parameters and parameter variations are usually *correlated.* For example, if the oxide thickness is near the upper limit of its tolerance interval, it will tend to push K' toward the lower limit and γ toward the upper limit of their corresponding tolerance intervals, as can be expected from Table 3.3. Also, if for a given process the absolute accuracy of γ is good, one can infer that the tolerance of the oxide thickness is narrow (since N_B and C'_{ox} are practically independent); thus, it can reasonably be expected that the absolute accuracy of K' will be good, too.

To take correlations into account, one can start from the tolerance of basic process parameters, such as oxide thickness and substrate doping concentration, and derive the resulting electrical parameters based on those [34]. Such a procedure can lead to sets of correlated extreme-case parameter combinations. For digital circuit design, e.g., one can derive extreme-case combinations corresponding to "slow" and "fast" devices. For analog circuits, though, what constitutes "extreme" cases may have to do with performance parameters other than speed (e.g., linearity or accuracy), depending on the application. Unfortunately, often no extreme-case parameter combinations specifically for analog design are available. Analog designers are thus forced to use "digital" worst-case parameter files, plus their own judgment, and constantly seek to use circuit techniques that result in performance which is robust against parameter tolerances and temperature variations.

For a given temperature, the variations of element values can be grouped into several categories:

1. *Lot-to-lot.* A "lot" of several wafers fabricated today can be markedly different from a lot fabricated using "the same" process a week or a year ago due to unpredictable variations in the fabrication conditions. The tolerances discussed above refer to such variations.

2. *Wafer-to-wafer* (in the same lot). For all wafers processed simultaneously in a single lot, fabrication conditions are similar, but not iden-

tical. Thus significant wafer-to-wafer variations still exist, although they are smaller than lot-to-lot variations.

3. *Chip-to-chip* (on the same wafer). These variations (*interdie variations*) are even smaller than the above. Still, though, the fabrication environment is not perfectly uniform over the wafer surface (e.g., furnace temperature or implant dose), and two "identical" devices on different chips of the same wafer can actually be significantly different, especially when the chips are not located close to each other on the wafer.

4. *Element-to-element* (on the same chip). This type of variation (*intradie variation*) is the smallest of all, especially if the two elements are located next to each other so that they undergo nearly identical fabrication processing. Thus, although the absolute values of their parameters cannot be predicted with any accuracy, the values can be expected to be nearly equal (*matched*). For example, two transistors next to each other on a chip taken from one lot can have threshold voltages of 0.760 and 0.763 V; if the two devices are chosen from a chip in another lot, the thresholds may turn out to be 0.852 and 0.856 V. In each case, the thresholds of the two devices match within a few millivolts. Two bipolar transistors next to each other with equal currents can have their V_{BE} values matched within 0.3 mV. Matching is very important in analog circuit design. Techniques to optimize device matching are discussed in Sec. 6.4.

The above observations can be extended to the performance of entire chips. Thus, the performance of two chips from different lots may be noticeably different; if the chips are from two wafers in the same lot, their performance could be closer. If they are from the same wafer, their performance can be even closer, especially if the two chips are located near each other on the wafer. Circuit design cannot rely on such observations, though. It should directly aim at producing correctly operating circuits on any wafer in any lot. Performance differences from lot to lot should be within the margins allowed by the specifications.

5.8 Chip Size and Yield

As with any manufacturing process, a number of things can go wrong during chip production. The ratio of the number of packaged chips that function correctly to the total number of chips one started with on the wafers is called the final *yield*. This can vary widely (from 0 to nearly 100 percent) and is affected by defects on the wafer, problems during fabrication, unreliable wire bonds, improper handling, etc. Here we will concentrate on the intermediate yield at the end of the wafer fabrication process. This quantity is intimately related to chip size, as illustrated in Fig. 5.28. We assume for the sake of this example that in both cases shown in the figure the same defects (indicated by dots) occur at the same locations, and that one defect per chip is enough to render that chip nonfunctional. The defects are supposed to be due to fabrication, independent of the nature of the circuits on the chips. In Fig. 5.28*a,* where a small chip size is used, the yield is 87/94 ≈ 93 percent. In Fig. 5.28*b,* where the chip size is large, the yield is zero.

The above simple example illustrates why chips cannot be made too large in a given technology. Given that the fabrication cost per (good) chip equals the wafer fabrication cost divided by the number of good chips, it becomes obvious that the cost per chip can grow dramatically as the chip size is increased. There is a constant drive to reduce the number of defects per wafer, thus increasing the yield for a given chip size. At the same time, a large amount of effort is devoted to making possible ever-smaller physical dimensions for the individual devices of which chips are composed. This scaling down of physical device dimensions, if done reliably, is desirable for the following reasons:

- The circuit speed increases, since the electrons, holes, and electric signals travel over shorter distances.

- For a given maximum chip size, more devices can be put on the chip, and thus more complex systems can be integrated.

- A given circuit can be made on a smaller chip, thus allowing more chips per wafer, increased yield, and decreased cost.

Of course, if the devices are made too small, one encounters individual device reliability problems. For a given technology, then, the process engineers decide on a minimum feature size (say, the gate length of a MOSFET), and circuit designers avoid making devices with fea-

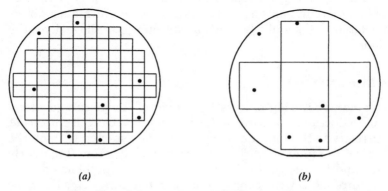

(a)   *(b)*

Figure 5.28 The effect of chip size on yield. The two wafers are assumed to have identical defects. (*a*) Small chip size; (*b*) large chip size.

tures smaller than that. Feature sizes of a fraction of a micrometer can be found on commercial chips. At the same time, chips over 1 cm^2 can be made with reasonable yield, thus making possible fully integrated VLSI systems with millions of transistors. Still, though, for a given circuit, designers pay special attention to making the chip area small, in order to maximize the number of good chips per wafer. This is especially true in high-volume products, in which the extra design time spent on minimizing chip area will be eventually amortized among a very large number of chips. What is an acceptable size depends, of course, on the type of circuit and the economics involved. A memory chip with an area of 1 cm^2 might be acceptable, whereas an integrated filter with an area of 5 mm^2 might be deemed unacceptably large (costly).* On the other hand, for very small circuits, the wafer fabrication cost per chip can be so low that the cost of the final product is dominated by the cost of the package; in such cases, little benefit will be derived from trying to make the circuit even smaller.

The biggest chip that can be envisioned with current fabrication techniques is an entire wafer [7]. *Wafer-scale integration* is currently in its infancy. This goal can become more realistic, though, as fabrication techniques continue to be perfected and as fault-tolerant systems are developed.

The complexity and speed possible in today's technology have made testing a critical issue, especially when the system is analog or mixed analog-digital. Sometimes, efforts to establish a good testing procedure can take longer than the efforts to design a chip. Testing has to be considered carefully and as early in the design process as possible. Otherwise, a large amount of effort can be spent on the design of a complex system, only to find out at the end that it is not adequately testable. Testing is also expensive; for some circuits, the cost involved in testing for even a few seconds is not justified. Testability considerations may dictate making available additional contacts to internal nodes (important for debugging a prototype) of a circuit, fabricating test structures along with it, or even creating self-testing circuits.

5.9   The Influence of Pads and Package

Although the internal speed in ICs can be very high, when fast signals must be exchanged with the external world, problems arise due to the capacitance and inductance of package pins and bonding wires. This problem is aggravated by the fact that complex integrated systems often require a lot of connections to the external world, thus making it

---

* The unit of 1 mil^2 is sometimes used for chip area. 1 mil = 0.001 in = 25.4 μm, and thus 1 mil^2 = 6.45×10^{-4} mm^2.

necessary to pack many pins close to one another. This can interfere with speed. Further, such circuits often dissipate a lot of heat, and the packaging must be such as to ensure adequate cooling. Another issue with packaging is the mechanical strain that it can cause on the chip, which can affect the chip's electrical performance. Many package types exist for various applications [53–55]. The development of improved packages is a field in itself.

A simple model for the parasitics that need to be considered is shown in Fig. 5.29, where the part corresponding to two neighboring terminals is included, together with the corresponding pads on the chip. The pads and the associated protection circuitry (Sec. 5.2.8) have parasitic capacitance to the substrate C_p, which is in series with substrate resistance R_b. The bonding wire offers some inductance (L_2), as do the leads and pins of the package (L_1). The close proximity of the leads and pins to each other, and of the bonding wires to each other, causes some parasitic capacitance (C_{ll}) as well as some mutual inductance (M_1 and M_2). There is parasitic capacitance (C_{lp}) between the bonding wires and the package's paddle, the latter being the large metal area on which the chip sits and to which it is attached and which is thus connected to the chip's substrate. Finally, there is a parasitic capacitance (C_B) between the package paddle and the ground plane of the printed-circuit board on which the package is soldered.

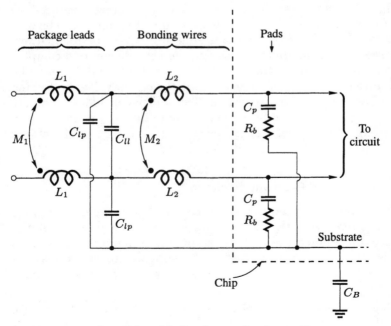

Figure 5.29   Lumped model of package and pad parasitics.

As an example, assuming a package used for RF applications in the 1 to 2 GHz range we may have L_1 and L_2 of 2 to 3 nH each, M_1 and M_2 about 1 nH, C_{ll} and C_{lp} about 100 fF, C_p about 400 fF, R_b a few hundred ohms, and C_B on the order of 1 pF. These values may vary depending on the type of package used and on the position of the leads within the package (the leads going to the corners of the package being longer than those going to the center of its sides). Packages for low-frequency circuits may have much larger parasitics (e.g., interpin capacitance of 4 pF and pin inductance of over 20 nH).

Ideally, if a good package is used, at the frequency of operation the inductances will have negligible impedance and can be replaced by short circuits. The substrate resistance can be neglected similarly at such frequencies, in comparison with the higher impedance of the capacitance in series with it. The effect of the package then reduces to that of capacitive loading, which can be easily considered during design. At high frequencies, though (i.e., at frequencies not far from the upper limit of the frequency range in which the package is useful), all parasitics should be considered, either to make sure that they do not interfere with proper operation or to take their effect into account. Note, for example, that any "ground" connection to the external world must go through a lead inductance, and thus at high frequencies it will not be a true ground. Sometimes several leads and bonding wires are connected in parallel, in an attempt to reduce their effective inductance and make possible a more effective ground or power supply connection.

At extremely high frequencies, even the model of Fig. 5.29 may be inadequate; more LC sections may have to be used, to attain a better approximation of the distributed nature of the parasitics. Also, the bonding wire and lead resistance may have to be taken into account. A correct model of the package to be used, for the frequency range intended, is one important piece of information that must be supplied by the manufacturer.

References

1. S. M. Sze, *VLSI Technology*, 2d ed., McGraw-Hill, New York, 1988.
2. W. S. Ruska, *Microelectronics Processing*, McGraw-Hill, New York, 1987.
3. S. K. Ghandi, *VLSI Fabrication Principles—Silicon and Gallium Arsenide*, 2d ed., Wiley, New York, 1994.
4. J. Y. Chen, *CMOS Devices and Technology for VLSI*, Prentice-Hall, Englewood Cliffs, NJ, 1990.
5. D. A. Antoniadis, "MOS Transistor Fabrication," in Y. P. Tsividis, *Operation and Modeling of the MOS Transistor*, McGraw-Hill, New York, 1987.
6. E. Demoulin, "Fabrication Technology of MOS IC's for Telecommunications," chap. 1 in Y. Tsividis and P. Antognetti, eds., *Design of MOS VLSI Circuits for Telecommunications*, Prentice-Hall, Englewood Cliffs, NJ, 1985.
7. S. K. Tewksbury, *Wafer Scale Integrated Systems: Implementation Issues*, Kluwer Academic Publishers, Boston, 1989.

8. D. J. Allstot and W. C. Black, Jr., "Technological Design Considerations for Mono-lithic Switched-Capacitor Filtering Systems," *Proceedings of the IEEE,* 71(8): 967–986, August 1983.

9. W. H. Hayt, Jr., and J. E. Kemmerly, *Engineering Circuit Analysis,* 2d ed., McGraw-Hill, New York, 1993.

10. J. McCreary, "Matching Properties, and Voltage and Temperature Dependence of MOS Capacitors," *IEEE Journal of Solid-State Circuits,* 16:608–616, December 1981.

11. J. B. Shyu, G. C. Temes, and K. Yao, "Random Errors in MOS Capacitors," *IEEE Journal of Solid-State Circuits,* 17: 1070–1076, December 1982.

12. J. B. Shyu, G. C. Temes, and F. Krummenacher, "Random Error Effects in Matched MOS Capacitors and Current Sources," *IEEE Journal of Solid-State Circuits,* 19: 948–955, December 1984.

13. A. T. Behr, M. C. Schneider, S. N. Filho, and C. G. Montoro, "Harmonic Distortion Caused by Capacitors Implemented with MOSFET Gates," *IEEE Journal of Solid-State Circuits,* 27: 1470–1475, October 1992.

14. M. Ghausi and J. Kelly, *Introduction to Distributed-Parameter Networks; with Applications to Integrated Circuits,* Holt, Rinehart and Winston, New York, 1968.

15. Y. Tsividis, M. Banu, and J. Khoury, "Continuous-Time MOSFET-C Filters in VLSI," *IEEE Journal of Solid-State Circuits,* 21: 15–30, February 1986.

16. I. Bahl and P. Bharta, *Microwave Solid-State Circuit Design,* Wiley, New York, 1988.

17. N. M. Nguyen and R. G. Meyer, "Si IC-Compatible Inductors and *LC* Passive Filters," *IEEE Journal of Solid-State Circuits,* 25: 1028–1031, August 1990.

18. K. Ashby, I. Koullias, W. Finley, J. Bastek, and S. Moinian, "High *Q* Inductors for Wireless Applications in a Complementary Silicon Bipolar Process," *Proceedings of the IEEE Bipolar/BiCMOS Circuits and Technology Meeting,* Minneapolis, MN, October 1994, pp. 179–182.

19. J. Y.-L. Chang, A. Abidi, and M. Gaitan, "Large Suspended Inductors on Silicon and Their Use in a 2-μm CMOS RF Amplifier," *IEEE Electron Device Letters,* 14: 246–248, 1993.

20. Y. Tsividis and R. W. Ulmer, "A CMOS Voltage Reference," *IEEE Journal of Solid-State Circuits,* 13: 774–778, 1978.

21. E. Vittoz, "MOS Transistors Operated in the Lateral Bipolar Mode and Their Application in CMOS Technology," *IEEE Journal of Solid-State Circuits,* 18: 273–279, June 1983.

22. A. A. Abidi, V. Comino, and T.-W. Pan, "The Potential of Using Parasitic Bipolars in CMOS for Analog Circuits," *Proceedings of the Bipolar Circuits and Technology Meeting,* Minneapolis, MN, September 1987, pp. 90–93.

23. E. A. Vittoz, "Micropower Techniques," chap. 3 in J. E. Franca and Y. Tsividis, eds., *Design of Analog-Digital VLSI Circuits for Telecommunications and Signal Processing,* Prentice-Hall, Englewood Cliffs, NJ, 1994.

24. M. Shoji, *CMOS Digital Circuit Technology,* Prentice-Hall, Englewood Cliffs, NJ, 1988.

25. J. P. Uyemura, *Fundamentals of MOS Digital Integrated Circuits,* Addison-Wesley, Reading, MA, 1988.

26. L. A. Glasser and D. W. Dobberpuhl, *The Design and Analysis of VLSI Circuits,* Addison-Wesley, Reading, MA, 1985.

27. N. Weste and K. Eshraghian, *Principles of CMOS VLSI Design,* Addison-Wesley, Reading, MA, 1993.

28. R. L. Troutman, *Latchup in CMOS Technology: The Problem and Its Cure,* Kluwer Academic Publishers, Boston, 1986.

29. D. B. Estreich and R. W. Dutton, "Modeling Latch-up in CMOS Integrated Circuits," *IEEE Transactions on CAD,* 1: 157–162, 1982.

30. A. R. Alvarez, ed., *BiCMOS Technology and Applications,* Kluwer Academic Publishers, Boston, 1989.

31. D. Stolta, R. Reuss, J. Ford, B. Cosentino, D. Monk, F. Shapiro, D. Lawey, and C. Dragon, "A BiCMOS 0.8 m Process with a Toolkit for Mixed-Mode Design," *Proceedings of the IEEE Custom Integrated Circuits Conference,* San Diego, CA, May 1993, pp. 24.2.1–24.2.4.

32. T. M. Liu, T.-Y. Chiu, and R. G. Schwartz, "High Performance BiCMOS Technology," *Proceedings of the IEEE Custom Integrated Circuits Conference,* San Diego, CA, May 1993, pp. 24.1.1–24.1.6.

33. D. J. Roulston, *Bipolar Semiconductor Devices,* McGraw-Hill, New York, 1990.
34. H. C. de Graaff and F. M. Kaassen, *Compact Transistor Modeling for Circuit Design,* Springer-Verlag, New York, 1990.
35. R. M. Warner, Jr., and B. L. Grung, *Semiconductor Device Electronics,* Holt, Rinehart and Winston, Philadelphia, PA, 1991.
36. U. Cilingiroglu, *Systematic Analysis of Bipolar and MOS Transistors,* Artech House, Boston, 1993.
37. S. M. Sze, *Physics of Semiconductor Devices,* Wiley, New York, 1981.
38. I. Getreu, *Modeling the Bipolar Transistor,* Tektronix Inc., Beaverton, OR, 1976.
39. P. R. Gray and R. G. Meyer, *Analysis and Design of Analog Integrated Circuits,* 3d ed., Wiley, New York, 1993.
40. Y. P. Tsividis, "Accurate Analysis of Temperature Effects in I_C-V_{BE} Characteristics with Application to Bandgap Reference Sources," *IEEE Journal of Solid-State Circuits,* 15: 1076–1084, December 1980.
41. B. Jayant Baliga, ed., *High Voltage Integrated Circuits,* IEEE Press, Piscataway, NJ, 1988.
42. P. Antognetti, ed., *Power Integrated Circuits—Physics, Design and Application,* McGraw-Hill, New York, 1986.
43. M. R. Haskard and I. C. May, *Analog VLSI Design—nMOS and CMOS,* Prentice-Hall, Australia, 1988.
44. S. D. Senturia, "Microsensors vs. ICs: A Study in Contrasts," *IEEE Circuits and Devices Magazine,* 6: 20–27, 1990.
45. R. F. Wolffenbuttel, "Integrated Micromechanical Sensors and Activators in Silicon," *Mechatronics I,* 4: 371–391, 1991.
46. K. T. V. Grattan, *Sensors—Technology, Systems, and Applications,* Adam Hilger, Bristol, England, 1992.
47. H. Baltes, "CMOS as Sensor Technology," *Sensors and Actuators A,* 37–38: 51–56, 1993.
48. R. F. Wolffenbuttel, "Fabrication Capability of Integrated Silicon Physical Sensors," *Sensors and Actuators A,* 41–42: 11–28, 1994.
49. S. K. Kai and V. K. Dham, "VLSI Electrically Erasable Programmable Read-Only Memory (EEPROM)," in N. G. Einspruch, ed., *VLSI Handbook,* vol. 13, Academic Press, New York, 1985.
50. C. Plett, M. Copeland, and R. Hadaway, "Continuous Time Filters Using Open Loop Tuneable Transconductance Amplifiers," *Proceedings of the 1986 International Symposium on Circuits and Systems,* San Jose, CA, pp. 245–248.
51. E. Sakinger and W. Guggenbuhl, "An Analog Trimming Circuit Based on a Floating-Gate Device," *IEEE Journal of Solid-State Circuits,* 23: 1437–1440, December 1988.
52. L. Carley, "Trimming Analog Circuits Using Floating Gate Analog MOS Memory," *Digest of 1989 International Solid-State Circuits Conference,* New York, pp. 202–203.
53. K. M. Striny, "Assembly Techniques and Packaging of VLSI Devices," chap. 13 in S. M. Sze, *VLSI Technology,* 2d ed., McGraw-Hill, New York, 1988.
54. R. R. Tummala and E. J. Rymaszewski, *Microelectronics Packaging Handbook*, Van Nostrand Reinhold, New York, 1989.
55. H. B. Bakoglu, *Circuits, Interconnections, and Packaging for VLSI,* Addison-Wesley, Reading, MA, 1990.

Problems

Note: Unless otherwise specified, all electrical parameters in the following problems are as in Examples 5.1 and 5.2.

5.1 A simple passive filter has a cutoff frequency f_c given by $f_c = 1/(2\pi RC)$, where R is the resistance of a poly resistor and C is the capacitance of a metal-poly capacitor. The nominal values of R and C are 10 kΩ and 50 pF, respectively, at room temperature. Find the worst-case (extreme) values of f_c at room temperature, assuming tolerances of ±20 percent for the resistor and ±10 percent for the capacitor. Also, find the worst-case values of the nominal f_c at 0 and 70°C, assuming temperature coefficients of 500 ppm/°C for the resistor and 25 ppm/°C for the capacitor.

5.2 The temperature coefficient of a capacitor C_1 is 25 ppm/°C; that of a neighboring capacitor C_2 is 23 ppm/°C. What is the temperature coefficient of the ratio C_1/C_2?

5.3 A well resistor (Fig. 5.8f) has a zero-bias sheet resistance of 1 kΩ/□ ± 20 percent, a temperature coefficient of 6000 ppm/°C, and a voltage coefficient of 10,000 ppm/V. The length and width tolerances are ±0.3 μm. When it is connected in a circuit, one end of the resistor is at 5 V and the other at 2 ± 0.2 V. The resistor is laid out with a nominal width of 5 μm. Find the maximum nominal length that will guarantee a resistor current of at least 100 μA, over the temperature range 0 to 70°C. (*Hint:* To use the given single voltage coefficient value, you will have to make a reasonable simplification appropriate to this worst-case design problem.)

5.4 (*a*) Derive the expressions for the small-signal current gain and the transition frequency f_T of a bipolar transistor (where f_T is defined as the frequency at which the small-signal current gain drops to 1), using the small-signal equivalent circuit of Fig. 5.21 with $C_{bs} = 0$, $g_\mu = 0$, and $r_{ex} = r_c = 0$ (see Sec. 5.3.2). In calculating the small-signal drain current, you can neglect its component through C_μ.

(*b*) Rederive the expression for the small-signal current gain, using the transadmittance y_m in Eq. (5.3) in lieu of the transconductance. Verify Fig. 5.24.

5.5 Consider the layout of an *npn* bipolar transistor in a BiCMOS process shown in Fig. 5.26. Assume the n^+ and p^+ strips are contacted over most of their length. Show that, approximately, the resistances in the device are inversely proportional to W_E, whereas the capacitances are proportional to W_E. At what W_E is the error involved in such an assumption maximum?

5.6 An *npn* bipolar transistor with $W_E = 20$ μm (Fig. 5.26), implemented by using the BiCMOS process of Example 5.2, is biased at a collector current of 600 μA, a collector-substrate voltage of 1 V, and a collector-base voltage of 1.5 V. Determine the values of all elements in its small-signal equivalent circuit (Fig. 5.21).

5.7 An *n*MOS transistor with nominal $W = 10$ μm, nominal $L = 3$ μm, nominal parameters as in Example 5.1, and tolerances equal to the typical ones given in Sec. 5.7 is biased with $V_{GS} = 2$ V, $V_{DS} = 4$ V, and $V_{SB} = 3$ V. Find the worst-case (maximum and minimum possible) values of the drain current.

6

Layout

6.1 Introduction

Before an integrated circuit can be fabricated, one must specify the geometric patterns associated with the devices in it. These patterns are "laid out" highly magnified on a computer terminal screen. The name *layout* refers both to the drawing process and to the resulting drawings.

A layout drawing is often presented as a single overlay of the various masks; basic masks for the CMOS process have been shown in Fig. 5.2 (top of each part). The result of overlaying all these masks can be confusing, so color coding is typically used to help differentiate between different masks (e.g., yellow for wells, green for active regions, red for poly, blue for metal, black for contact windows, etc.). In this book we will use a simpler style for layout, which will be illustrated shortly. From this it will be easy to produce the layout of each mask for a given technology.

A number of *layout rules* must be followed during the layout process. These rules are devised for each fabrication process, from considerations such as reliability and immunity to undesirable effects (e.g., breakdown and latch-up). Examples of possible rules are "poly regions should be not narrower than 1 μm," and "wells should be no closer to each other than 5 μm." By the time all layers and their proximity or overlap in various combinations are considered, one ends up with several dozen such rules. The rules vary, depending on the fabrication facility and type of process used. A simplified set of rules will be given as an example later.

Layout is accomplished with the help of computer aids. The user can specify every line or shape used to form the numerous geometric patterns. Some layout aids can be used to draw entire devices with proper geometry, with the user only having to specify certain key dimensions (such as channel width and length) and the approximate location of the device. Entire layouts of devices or subcircuits that are used often can be stored in memory, and they can be recalled and added to the layout. The routing of interconnections between such blocks can be accomplished automatically in some cases, by using programs called *routers*. Computer aids have made possible the layout of large VLSI chips in reasonably short time, a task which would have been impossible otherwise.

Following extensive verification and design iterations, the information contained in the layout (in the form of a computer file) is sent to the fabrication facility, where it is used to automatically produce the masks for the fabrication process. At times, the layout information is submitted to an intermediary organization, whose function is to interface with a number of potential chip makers. Each chip maker may make slight modifications to certain feature sizes on the layout, appropriate to its own fabrication process.

In our discussion of layout in this chapter, we begin with considerations that apply to both digital and analog circuits. The material following these general considerations focuses on issues that are of particular importance for analog and mixed analog-digital circuits: device matching and interference control.

6.2   Relation of Fabricated Transistors to Layout

The actual values of the channel width and length after fabrication are different from their values on the mask. Several reasons for this were mentioned in our description of the CMOS fabrication process in Sec. 5.2.1. Thus, we have

$$W = W_{\text{mask}} - \Delta W \tag{6.1}$$

with W the final width after fabrication, W_{mask} the channel width on the mask, and ΔW the width reduction due to channel "encroachment," leading to the "bird's-beak" effect (Sec. 5.2.1) and to effects related to ion implantation in the channel-stop regions. Typically ΔW is several tenths of a micrometer. Similarly, we have

$$L = L_{\text{mask}} - \Delta L \tag{6.2}$$

where L is the final, actual length of the channel (often called the *effective channel length*), L_{mask} is the channel length on the mask, and ΔL is the total channel length reduction. Considering the individual causes of this reduction, we can write the above equation as

$$L = L_{\text{mask}} - l_r - 2l_{\text{lat}} \tag{6.3}$$

where l_r (positive or negative, depending on the technology) is the size reduction during the formation of the poly gates and l_{lat} is the lateral diffusion length on each side. These quantities can typically be up to a few tenths of a micrometer each.

In our layouts, we will draw the poly gates of transistors as we would like them to appear finally on the chip. In other words, we will not need to concern ourselves with l_r and ΔW; we will assume that the manufacturer, based on an estimate of these quantities, will use appropriate values for W_{mask} and L_{mask} so that *after* fabrication the devices will end up approximately as we have intended.* However, we *will* have to take into account the effect of lateral diffusion; in other words, if we want a channel length L as shown in Fig. 6.1b, we will have to lay out a device with a poly gate as shown in Fig. 6.1a. The horizontal broken lines in

---

* While this is a convenient assumption, we should mention that it does not reflect universal practice; in certain design environments, it is required that the effects of ΔL and ΔW be completely taken into account during the layout process.

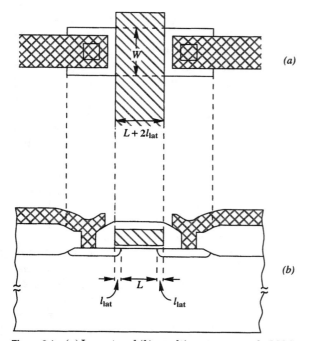

Figure 6.1 (*a*) Layout and (*b*) resulting structure of a MOS transistor, showing the reduction in channel length due to the lateral diffusion (other reductions are not included here—see text).

this figure show the boundary of the thin oxide. We should note though that even after the corrections l_r, l_{lat}, and ΔW are taken into account, there will still exist an uncertainty about the final channel length and width of the fabricated devices due to fabrication tolerances. Such uncertainties can be as large as ±0.2 μm (lot to lot). Throughout this book, W and L stand for the *actual geometric* values of the channel width and length of the fabricated devices, as shown in Fig. 6.1.

6.3   Transistor Geometry and Layout

We now show simple MOS device layouts for several cases. Figure 6.2a shows a device with $W/L < 1$. We are assuming that the minimum W allowed in the fabrication process is used. To achieve the desired $W/L < 1$, L is made large as shown. Narrow-channel effects (Sec. 3.8) or poor matching to other devices (Sec. 6.4) may make this device poorly suited for certain circuits. To reduce such effects, one can use the device in Fig. 6.2b, which has a nonminimum W but maintains the same W/L ratio (in this discussion lateral diffusion is neglected for simplicity, unless it is specifically mentioned; if its magnitude is known, it can be taken into account as explained in Sec. 6.2). The penalty is, of course, increased chip area and increased capacitances (particularly intrinsic ones).

In Fig. 6.2c we show a device with $W/L > 1$. The diffusion regions are contacted over most of their width, to keep the source and drain resistance low. Notice that multiple contact windows are used, rather than one long window. By having only a single window size and selecting this size appropriately, the process can be optimized to provide good-quality contacts. In Fig. 6.2c we assume that the channel length is the minimum allowed in the process; the channel width is as large as needed to implement the needed W/L value. In many cases in analog circuits, a minimum-value L is not desirable, because it can lead to excessive drain-source small-signal conductance and other short-channel effects (Sec. 3.8); also, the matching between minimum-length devices can be poor (Sec. 6.4). Figure 6.2d shows a device with a longer channel than in Fig. 6.2c, but with the same W/L. Increased chip area and increased capacitances are again the resulting penalty.

Large W/L ratios can be implemented as in Fig. 6.3a. As indicated in Fig. 6.3b, this device consists of four subdevices in parallel; in this example, each of these has the W/L ratio of the device in Fig. 6.2c. Thus, the effective W/L ratio in Fig. 6.3a is 4 times larger than that in Fig. 6.2c. The diffusion strips, with the exception of the two outer ones, are shared by two subdevices each. This reduces the required area, in comparison to a device laid out in a straightforward manner and with the same total W. The junction capacitances are also reduced. One reason for this is that most bottom-wall areas are shared by two subde-

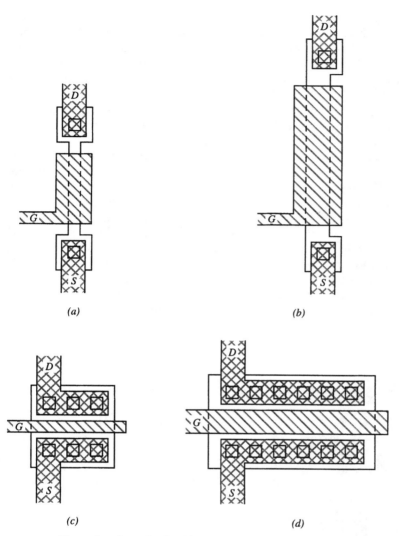

Figure 6.2 Examples of standardized layout. The various cases are discussed in the text.

vices each. Another reason has to do with the sidewalls. The external outline of the device defines the *active area* (Sec. 5.2.1) and is surrounded by a channel-stop region as usual. Most of the diffusion periphery in Fig. 6.3a (especially that of the drain) is inside the active area, and it does not come in contact with channel-stop regions; thus it does not exhibit the high sidewall capacitance of more conventional structures.

Note that if the device in Fig. 6.2c has been well characterized, then
the one in Fig. 6.3a is well characterized also (except for junction
capacitances). The simplicity, modularity, and predictability of devices
like the ones in Figs. 6.2c, 6.2d, and 6.3a make them attractive despite
the associated area penalty, and many designers use such devices
extensively.

In certain cases, contacts to source or drain regions can be avoided
altogether. An example is the layout of the "cascode" circuit shown in
Fig. 6.4a. Here no external connection is needed to point X. By laying
out the circuit as in Fig. 6.4b, a common strip is used for both the drain
of the bottom device and the source of the top device. Since no contacts
are needed along this strip, it can be made narrow. This makes possible
a low parasitic junction capacitance at point X and a compact layout.

In certain settings, the modularity afforded by the layout style illustrated in Figs. 6.2 and 6.3 is abandoned in favor of totally "handcrafted"
layout. Examples are shown in Fig. 6.5. In all cases in this figure, multiple contacts are avoided in order to minimize the junction areas
(which have to accommodate clearance around each contact; see Example 6.1). This results in smaller total chip area as well as smaller junction capacitances. For Fig. 6.5a, the W/L value is the same as for Fig.

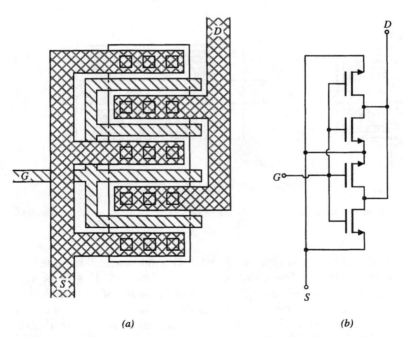

(a)   (b)

Figure 6.3 Transistor with large W/L. (a) Layout; (b) circuit diagram showing
subdevices and their connection.

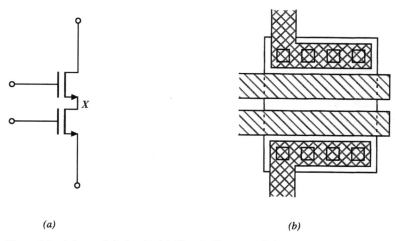

(a)   (b)

Figure 6.4   A "cascode" circuit. (a) Circuit diagram; (b) layout.

6.2c. However, here the source and drain resistances can be large, and undesirable voltage drops can develop across them if the current is significant. Since the resistances are distributed along the channel width, the channel voltage will be gradually decreasing below the externally applied drain-source voltage, as one moves away from the contacts. This effect is difficult to model. The use of silicides for the source and drain regions reduces, but does not eliminate, this problem.

In Fig. 6.5b we show a device with a much larger W/L; the idea behind this layout is the same as that for Fig. 6.3a. The I-V characteristics of devices such as this are usually assumed to be the same as those of devices with rectangular channels with an appropriate L and W. The perpendicular distance between drain and source is used for L, and an estimate is used for the effective W (Prob. 6.2). However, this is only an approximation; strictly speaking, it may not be possible to define an equivalent W and L, since the "corner" regions may introduce effects that cannot be modeled by a simple adjustment of those quantities. Devices like the one in Fig. 6.5b are usually not well characterized, and they may not match as well as simple devices. Also, the problem with source and drain resistances mentioned above is present in these devices, too. Thus, discretion is required in their use.

The device of Fig. 6.5b makes possible a compact layout due to the fact that the internal drain "fingers" come in contact with the channel on three sides. Extending this idea to four sides, we have the device of Fig. 6.5c. To avoid the uncertainties associated with the corners, sometimes (rarely) circular geometries are used for the gate and drain. Note that the drain is wholly within the active area and thus does not come into contact with the channel-stop region. Its sidewall capacitance is

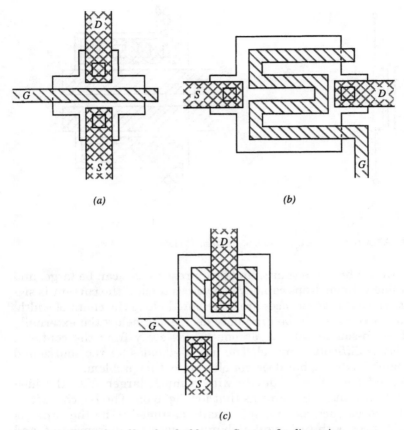

(a)   (b)

(c)

Figure 6.5   Examples of handcrafted layout. See text for discussion.

thus low. Also, since the gate is wholly within the active area, it is not adjacent to a "bird's beak" region (Sec. 5.2.1). The idea behind Fig. 6.5c can also be extended to very large W/L ratios by using multiple drain squares (see Prob. 6.3 and Fig. 6.27).

Handcrafted layouts like those in Fig. 6.5 may be used in cases where the drain and source resistances do not cause problems. For example, in micropower circuits the associated voltage drops can be negligible; also, in such cases the parasitic RC products due to the drain and source areas may not be a concern, since device speed at low currents is low anyway (Sec. 4.5). The rather poor modeling and characterization of such structures, though, can make their use problematic except in noncritical cases. In general, it is better to stick to the simpler layout style of Figs. 6.2 and 6.3, unless the designer has a lot of experience and the extra time spent on handcrafting, and on making sure the resulting circuit works properly, can be justified (see Sec. 6.6).

Bipolar *npn* devices in the BiCMOS processes discussed in Sec. 5.3 can be laid out as shown in Fig. 6.6*a;* a simplified form, chosen for clarity, is shown. Details can differ significantly, depending on the particular fabrication technology used. If the emitter area must be increased to accommodate a large current, one could increase W_E, leaving the horizontal dimensions unchanged at their minimum value; this would keep the parasitic series resistance low and would allow the rough scaling of electrical parameters mentioned in Sec. 5.3. If the needed emitter area is very large, though, a better layout involves multiple strips; such a layout can be done in various ways, one example being shown in Fig. 6.6*b*. Here the effective emitter area is the sum of the individual subemitter areas. The resistances from each base strip to the region under each emitter are in parallel, and the resulting effective base resistance is lower than what would be achieved if a straightforward layout with the same total emitter area were used. This is important for high-frequency work, since it achieves a high f_m, as discussed in Sec. 5.3.2. It is important that all emitter strips see the same V_{BE} value, as small differences in this quantity can cause large differences in the strip currents. This should be kept in mind when connecting the strips together. In power devices, further considerations are necessary, due to the heat generated by such devices. The emitter strips can be at different temperatures, which will make their currents different. In addition, an increase in a current because of heat can cause further heating, which can cause a further current increase, etc. (thermal runaway). To reduce these effects, sometimes small resistors are used in series with each emitter strip ("ballast resistors"); this, of course, affects the transistor performance (e.g., it reduces the effective transconductance).

The detailed layout of devices and circuits requires knowledge of a set of layout rules, set by device engineers and based on considerations of reliability, latch-up, immunity, etc. [1]. An example of such rules is given now.

Example 6.1 Simplified layout rules The rules given below are meant to be as simple as possible, as required for quick, rough estimates during circuit design. Full-fledged layout rules for industrial processes can be much more complicated, with several dozen rules being common for a given process. The layout rules given below can be used in conjunction with our example CMOS and BiCMOS 1-μm processes, the electrical parameters of which were given in Secs. 5.2 and 5.3 (Examples 5.1 and 5.2). All dimensions given are horizontal dimensions (looking at the device from the top, as in Fig. 6.2); the vertical dimensions (e.g., junction depths) are set by device and fabrication engineers, and the circuit designer has no control over them.

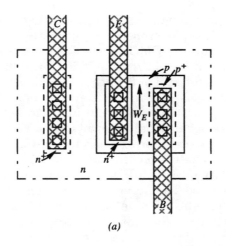

(a)

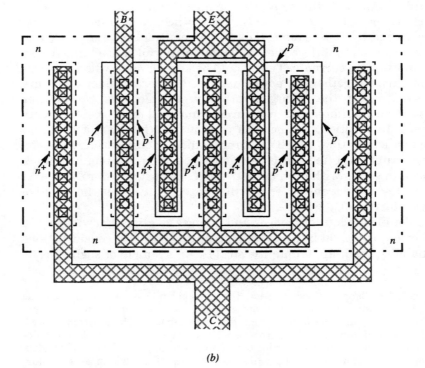

(b)

Figure 6.6 Bipolar transistor layout. (*a*) Simple layout; (*b*) multistrip layout.

Active areas*: At least 1.5 μm wide, spaced at least 2 μm apart.

Poly: At least 1 μm wide; should extend on either side of a transistor channel by 1 μm; at least 1-μm spacing between poly regions, or between poly and unrelated active areas.

Wells: At least 5 μm wide; spaced at least 5 μm apart; at least 3-μm spacing between a well edge and active areas inside or outside the well.†

Contact windows: 1-μm by 1-μm squares, with the region being contacted extending by at least 1 μm all around the contact window. Contacts should be spaced at least 1 μm apart, and should be at least 1 μm away from transistor channels.

Metal: At least 2 μm wide, spaced at least 2 μm apart. The metal should overlap each contact window completely and extend all around it by at least 0.5 μm.

Bonding pads: 125-μm by 125-μm square, spaced 25 μm apart. Pads should be at least 25 μm away from all circuits except those used for protection circuitry (Sec. 5.2.8).

Capacitors: Only metal-poly type is allowed (metal and poly are separated by a specially formed capacitor oxide), with top plate dimensions at least 2 μm by 2 μm. Follow "poly" and "metal" rules. For good matching between capacitors, use at least 25-μm by 25-μm top plates (Sec. 6.4), and extend bottom plates outside top plates by 3 μm all around the periphery.

Resistors: Follow the rules given above for the layer from which the resistor is to be made (Sec. 5.2.3). Widths larger than the minimum given should be used if good matching is desired (Sec. 6.4). Widths larger than the minimum may also have to be used for resistors forming *pn* junctions with their environment, if a low-voltage coefficient is necessary (especially for well resistors—see Sec. 5.2.3).

p (Base) diffusion (BiCMOS only): At least 1.5 μm wide; spaced at least 3 μm from external n^+, p^+, or p diffusion regions; should extend all around internal n^+ and p^+ regions by 1 μm; should be at least 5 μm away from well edges.

The application of rules such as these to the layout of digital circuits is discussed elsewhere [2–4]. The layout of analog circuits is illustrated in the following examples.

Example 6.2   Layout of a simple circuit.   Lay out the circuit of Fig. 6.7a (a differential amplifier [5–10]), using the layout rules of Example 6.1. Assume that the lateral diffusion extent l_{lat} is 0.25 μm.

solution   The channel lengths should be laid out 2×0.25 μm longer, to account for subsequent reduction due to lateral diffusion, as suggested by Fig. 6.1 and the associated discussion (recall our assumption that l_r and ΔW corrections, discussed in Sec. 6.2, will be carried out separately before the mask is made). The layout can be conveniently drawn on a square grid, as shown in Fig. 6.7b. The grid size corresponds to a "unit," of which all layout rule numbers are multiples; in our case, this "unit" is 1 μm. The reader can check that all layout rules given

---

* Areas where n^+, p^+, or transistor channels are to be formed; see Sec. 5.2.1.

† An exception to this is an n^+ contact inside the well; since such a contact is electrically connected to the well, no spacing is required between it and the well edge.

above are satisfied by the layout in Fig. 6.7b. This layout is a most straightforward one; no attempt has been made to minimize the layout area. Also, we have not used any special techniques to optimize the matching between the two middle, or the two bottom, devices. Such techniques will be discussed in Sec. 6.4. We have assumed that the power supply and ground buses also feed other circuits on the same chip. They are laid out wide not only because they may have to carry a large dc current, but also to reduce possible interference (Sec. 6.5).

Example 6.3   Stacked layout.   Lay out the circuit of Fig. 6.8a, using the layout rules of Example 6.1 and assuming a lateral diffusion of 0.25 μm.

solution   The drain channel length should be 2.5 μm + 2(0.25 μm) = 3 μm. A possible layout is shown in Fig. 6.8b, [11, 12] and can be explained with the help of Fig. 6.8c. Subdevices M1A and M1B share a drain, and together they form M1; similarly for M2A and M2B, which form M2. Device M3 is composed of M3A and M3B in parallel. The source of M1B serves also as the drain of M3A, and the source of M2B serves as the drain of M3B. Compared to a nonstacked layout, the layout of Fig. 6.8b achieves smaller area and smaller parasitic capacitances, as already discussed. Even more area can be saved in a two-level metal process; in such a case, the metal lines running outside the active areas can be eliminated, and the various source and drain areas can be connected by second-level metal lines running above the active area [11] (provided that matching is not adversely affected—see Sec. 6.4.4). The second-level metal lines are separated from the first-level metal by thick oxide, and thus the two levels are not short-circuited (except, of course, where the two levels are intentionally connected through vias). Other examples of stacked layouts can be found in the references [11, 12].

Example 6.4   Operational amplifier layout.   Figure 6.9a shows the schematic of an operational amplifier [5–10]. A layout for this circuit is shown in Fig. 6.9b.

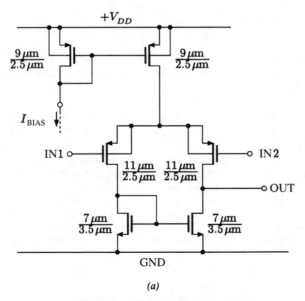

(a)

Figure 6.7   A differential amplifier. (a) Schematic.

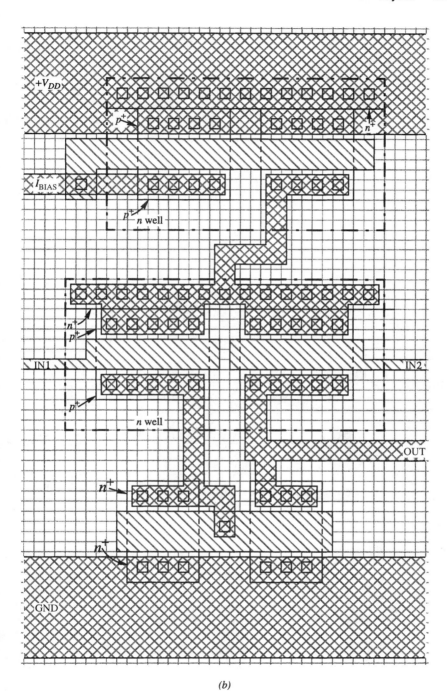

(b)

Figure 6.7 (*Continued*) A differential amplifier. (*b*) Simple layout.

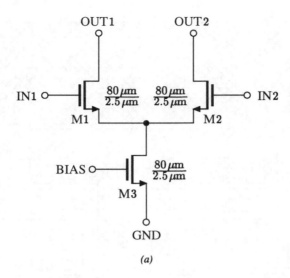

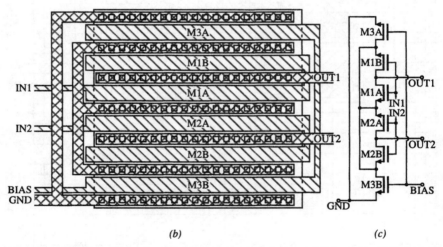

Figure 6.8   A differential pair with a current source device. (*a*) Schematic; (*b*) "stacked" layout; (*c*) circuit diagram showing subdevices and their connection.

Here the stacked layout style [11, 12] discussed in Example 6.3 is used extensively. The V_{DD}, GND, and BIAS lines run the entire width of the op amp cell, and can be extended horizontally to connect to other cells if desired.

Layout dimensions must be taken into account when one is preparing an input file for circuit simulation; an example is given in App. C.

The lambda rules.   In simplified layout rules [13], one often uses a "least common denominator," which is considered to be the "funda-

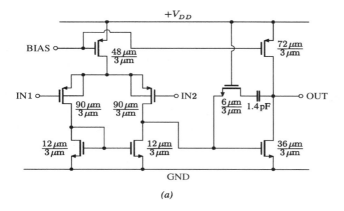

(a)

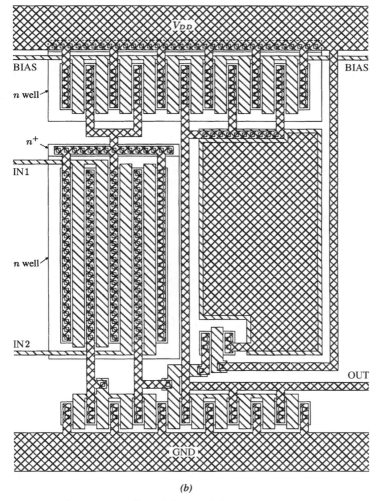

(b)

Figure 6.9 An operational amplifier. (*a*) Schematic; (*b*) layout.

mental resolution of the process" and is denoted by λ (not to be confused with the same symbol when it is used sometimes to represent the inverse of what we have called V_A in Sec. 3.2.1). All layout rules are then given in terms of λ; for example, the poly layer must be at least 2λ wide, the metal must overlap the contact windows by at least λ, etc. ($\lambda = 0.5$ µm in the layout rules of Example 6.1). The idea here is to make the layout rules relative, so that the same design can be fabricated by different manufacturers using similar processes with somewhat different values of λ. Also, if a fabrication process is improved and dimensions can be reduced, one hopes to be able to simply specify a new value for λ, leaving all *relative* dimensions the same; thus one hopes to use the layouts of existing designs to produce scaled-down circuits fabricated with a new process. While the idea is convenient and attractive in certain settings (e.g., university research), state-of-the-art commercial chips are not usually made this way, especially analog and mixed analog-digital ones. Usually when a new process is developed, a new set of rules is devised, optimized for that particular process. In this way better performance, better packing density, and/or better reliability can be achieved. Such individual customization of layout rules becomes especially important in modern processes with channel lengths of submicron dimensions.

6.4 Layout for Device Matching and Precision Parameter Ratios

The ability to match devices located close to each other on a chip is extremely important for analog IC design and has been widely exploited. Circuits have been devised whose performance depends critically not on absolute values of quantities, but only on certain quantities being equal. More generally, properly designed analog circuits often rely on *ratios* of like quantities (capacitance ratios, resistance ratios, current ratios, etc.) [5–10]. The accuracy of such ratios can be high if proper layout techniques are used [5, 6, 12, 14–24]. We will now discuss several such techniques.

6.4.1 Capacitor layout

Top views of capacitors have been given in Figs. 5.3a and 5.5. Assume that it is desired to make two identical capacitors and that their top plates are laid out very small (as seen from the top, looking down on the plates), as shown by the broken lines in Fig. 6.10a (the bottom plates are not shown, for simplicity). After fabrication processing, the top plates of the capacitors can end up looking as shown by the solid lines. The actual edges are inside the intended areas by an average distance that is not precisely known but is practically the same in both capacitors, and there is a small-scale random fluctuation of the edge position. The ratio of the two capacitor values can obviously be significantly different from unity. Assume now that the two capacitors were laid out to be larger, as shown in Fig. 6.10b. Now the small-scale randomness of the edges contributes a smaller relative error, and thus the ratio accuracy is better. The effect of small-scale random oxide thickness variations also becomes less severe when plate dimensions are increased, since it tends to average out over their larger area.

Thus, when elements must be critically matched, it often pays to use for them not the smallest dimensions possible for a given fabrication process, but larger ones. This holds for capacitor plate dimensions as well as resistor and transistor width and length. Of course, one should not overdo this; if the two plates in Fig. 6.10b are made *too* large, not only will the resulting chip area be excessive, but also there is a chance that parts of the two capacitors located at opposite ends will undergo different fabrication processing, resulting, say, in slightly different oxide thickness or different average location of the actual edges with respect to the intended location; in this case the ratio accuracy can deteriorate.

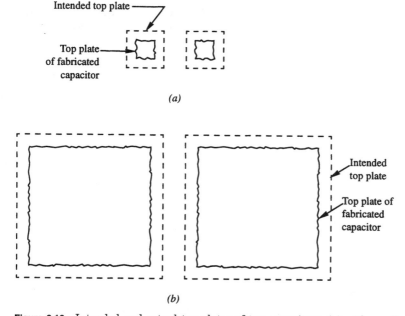

(a)

(b)

Figure 6.10   Intended and actual top plates of two capacitors, (a) with small dimensions and (b) with large dimensions. Bottom phases are not shown for simplicity.

Let us now consider the systematic error Δl in the edge location. We assume that this error is the same around the periphery of each capacitor. For simplicity, we also assume that no small-scale random fluctuations are present and that there is no oxide thickness variation across the chip. For a rectangular plate such as shown in Fig. 6.11 and using the symbols defined there, we see that if Δl is much smaller than the plate dimensions, we have an area error of $\Delta A \approx P\,\Delta l$. The relative error in the plate area (and thus in the capacitance, under our simplifying assumptions) is $\Delta A/A \approx (P/A)\,\Delta l$. It follows that for capacitors of different dimensions the relative error will be the same if the *perimeter/area ratio P/A* is the same. This is a very desirable situation since if two capacitor values exhibit the same *relative* error (e.g., if they are both smaller by 5 percent), the *ratio* of the capacitances will not be in error.

To illustrate the above observation, assume that it is desired to make two capacitors C_1 and C_2, with $C_2/C_1 = \eta$. Then the layout in Fig. 6.12a (where the plates are originally laid out with an area ratio of η) can give poor results, since the perimeter/area ratio is not the same in both (in drawing the figure, $\eta = 1.5$ is assumed). This problem can be corrected by changing the shape of C_2 into a rectangle of the same area, but with an aspect ratio such that its perimeter is η times that of C_1. The required dimensions can easily be shown to be those in Fig. 6.12b (Prob. 6.8). This technique can be generalized to different shapes (Prob. 6.9). It is mostly used when $\eta < 2$. Other techniques sometimes employed to maintain a constant perimeter/area ratio include the use of notches or holes in the top plate.

It is easy to see that the same strategy of maintaining constant perimeter/area ratio works for minimizing the effect of errors due to

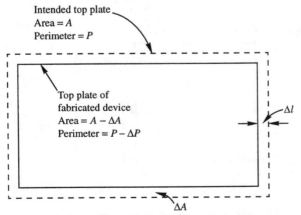

Figure 6.11 Intended and actual top plate of a rectangular plate capacitor, and definition of associated quantities.

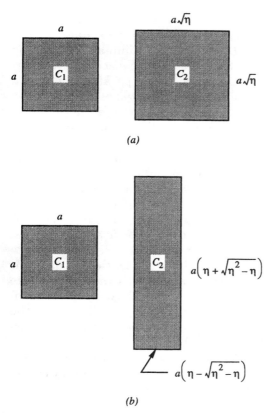

Figure 6.12 (*a*) A poor way and (*b*) a better way to implement a capacitance ratio $C_2/C_1 = \eta$.

the *fringing* field lines along the perimeter of the capacitor top plates (these are the field lines that escape outside the area immediately below the top plate). These lines contribute some extra capacitance, which will be in proportion to the main capacitance if the above strategy is used; thus, the ratio of the capacitances will not be affected.

A widely used approach for implementing integer capacitance ratios is to implement the larger one out of parallel-connected cells, each of which is identical to the smaller capacitor. Consider first the possibility illustrated in Fig. 6.13*a*, where we assume that $C_2/C_1 = 2$ is desired. Here, apart from random fluctuations, and in the absence of long-range effects, all three areas will be off by the same amount after fabrication; by connecting two of these capacitors in parallel as indicated, a ratio C_2/C_1 of very close to 2 can be achieved. The ratio accuracy can, however, be inadequate if there is a variation in capacitance per unit area with position, especially if it is in the horizontal direction on the page in Fig. 6.13a. Such variation can be caused, for example, by an oxide thickness gradient. Assuming the resulting capacitance variation is linear with position (Prob. 6.10), a way to eliminate its effect on the capacitance ratio is shown in Fig. 6.13b. The capacitors are interleaved, and the two at the ends are connected in parallel to form C_2. Although the individual values are not equal, the *sum* of the two capacitances forming C_2 is equal to $2C_1$. It is also easy to see that an oxide thickness gradient in the vertical direction will have no effect on the capacitance ratio accuracy.

As is obvious by now, to match capacitors, it makes sense to make everything identical, so that all capacitors are affected by the same factors in the same way. This is almost the case, for example, for Fig. 6.13b, except for one factor: The *environment* of the unit capacitors is not identical. Thus, the middle capacitor sees capacitors on both its left and right, whereas the others see a capacitor only on one of their sides. Even lacking the details of the fabrication process, it is reasonable to expect that the fabrication conditions will not be identical in different environments. This can happen, for example, with the etching rate during processing. To make the environments of all capacitors as identical as possible, dummy capacitors are often used in addition to the actual capacitors, at a distance equal to that between the actual capacitors themselves, as shown in Fig. 6.13c. Although it is best to use full-fledged dummy capacitors [19], often narrower strips are used to save area. In some settings it is recommended to ground the dummy capacitors, to avoid charge build-up during fabrication.

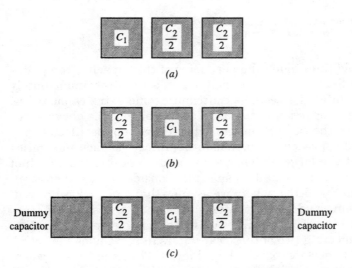

Figure 6.13 Implementing a ratio $C_2/C_1 = 2$ by using unit capacitors. The various cases are discussed in the text.

The above techniques improve the ratio accuracy dramatically, but still cannot do anything about the random edge fluctuations all along the perimeter (Fig. 6.10). The choice of square shapes provides the smallest perimeter/area ratio (compared to other practical, rectangular shapes) and thus minimizes the effect of random fluctuations. The reader may intuitively see why a nonsquare rectangular geometry will increase the effect of random fluctuations, by imagining a very elongated rectangle, in the extreme case that two sides almost touch. If the random fluctuations of these sides (always assumed to be rough) are visualized touching, it becomes evident that the accuracy will be totally unacceptable.

The splitting of capacitors into smaller, identical square capacitors (commonly called *unit capacitors*) is standard practice. The minimum allowed value of a unit capacitor is decided based on technological considerations and experiments, so that it will provide an acceptable ratio accuracy for the application at hand. Typical unit capacitors are 20 to 40 μm on a side. To reduce the effect of small-scale variations in the oxide thickness (caused, for example, by poly "bumpiness"), the capacitor oxide is usually made thicker than the gate oxide (e.g., 400 Å). The capacitance of a unit capacitor is typically 0.2 to 1 pF.

The use of unit capacitors can be extended to simple noninteger ratios, as illustrated in the following example.

Example 6.5 Lay out two capacitors C_1 and C_2 such that $C_2/C_1 = 1.5$.

solution Noting that $1.5 = \frac{3}{2}$, we use three unit capacitors in parallel for C_2 and two unit capacitors in parallel for C_1.

If C_2/C_1 cannot be written as a ratio of two small integers, as in the above example, additional considerations may be necessary. This is illustrated below.

Example 6.6 Lay out two capacitors C_1 and C_2 with a ratio $C_2/C_1 = 10.1$.

solution If we were to continue in the spirit of Example 6.5, we would have to write $10.1 = 101/10$ and use 101 unit capacitors for C_2 and 10 unit capacitors for C_1. This would occupy a large chip area and might even present problems for the circuits driving those capacitors. A better solution is to note that $10.1 = (9 \times 1 + 1.1)/1$. Thus, for C_1 we use one unit capacitor; C_2 will be formed of 10 capacitors, 9 of which will be identical unit capacitors and the tenth will be slightly augmented to have an area equal to 1.1 times that of a unit capacitor. It is important, though, to choose the shape of this capacitor so that its perimeter/area ratio is the same as that of the unit capacitors, for reasons already discussed above. This technique is widely used and provides satisfactory accuracy in most applications.

When several unit capacitors are to be laid out, they are sometimes placed symmetrically, as illustrated in Fig. 6.13c. This technique can be extended to whole arrays of capacitors which must maintain accurate ratios to each other. The resulting layout is made symmetric in both dimensions as much as possible and is referred to as a *common-centroid geometry* layout. It is commonly used for large capacitor arrays employed in A/D and D/A converters. Over the significant distances such arrays can span, large-scale effects such as oxide thickness gradients can impact ratio accuracy; a common-centroid layout will cancel such effects to a first order. However, such geometries will require extra wiring. In some instances, the associated complicated layout, extra chip area, and parasitics are undesirable, so this approach is not always taken, especially when the capacitor array size is not very large.

In the above discussion we have been showing only the top plate of each capacitor, for simplicity. A complete layout of a capacitor may look like that in Fig. 6.14a. The bottom plate is larger, so that the capacitance is defined by the area immediately below the top plate. However, some extra capacitance exists due to the curved, fringing electric field lines falling slightly outside this area. This extra capacitance is the same for all unit capacitors (we ignore random variations), so it does not affect the capacitance ratio. Consider, however, the narrow top-plate terminal "neck," which may be used for connection to other unit capacitors. If a mask misalignment occurs, the situation may look like that in Fig. 6.14b. Clearly, the parasitic capacitance contributed by this region will now be different, and depending on the layout it can affect the capacitance ratio accuracy. This problem can be corrected by laying out the unit capacitor as in Fig. 6.14c. If a misalignment is present, as shown in Fig. 6.14d, the total capacitance will be unchanged. The practice of using connecting necks, as shown in Fig. 6.14 (and in Figs. 6.15

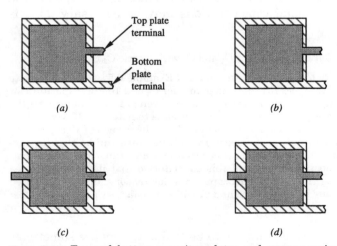

Figure 6.14 Top and bottom capacitor plates and narrow regions used for connecting them to a circuit. (a) Poor layout and (b) resulting position after misalignment; (c) correct layout and (d) resulting position after misalignment.

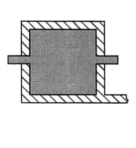

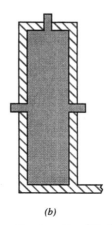

(a)   *(b)*

Figure 6.15   Layout of capacitors with a capacitance ratio of ⅔, the total parasitic capacitance of the narrow connection necks is proportional to the main capacitance.

to 6.17), is avoided in some technologies which do not allow a poly 2 line to run over a poly 1 edge for reliability reasons.

The contribution of the connecting necks (leaving now aside the possibility of misalignment) should also be taken into account for nonunit capacitors. Figure 6.15 shows the layout of a unit capacitor and a nonunit one with capacitance equal to 1.5 units. Since the nonunit capacitor has three necks and the unit capacitor has two, the ratio of the parasitics contributed by the necks is 3:2 = 1.5, just as for the main capacitances. Thus, the presence of the necks does not cause the capacitance ratio to deteriorate (assuming no vertical misalignment takes place). For other nonunit-to-unit capacitance ratios, though (e.g., 1.2 or 1.7), one cannot do an exact job since the number of necks must be an integer. For example, for a ratio of 1.7 one would still use three necks (and may consider increasing the width of one of them slightly).

The narrow neck of the bottom plates in Fig. 6.15 will present some moderate resistance (e.g., 100 Ω between two neighboring bottom plates). If such resistance cannot be allowed in the application at hand, the width of this neck can be increased. Two wide necks per side of bottom plate are sometimes used. In some cases, no separate bottom plates are used for each unit capacitor. Instead, a single bottom plate is used, large enough to accommodate all unit capacitor top plates above it. Whether this is appropriate or not will depend on the information provided by fabrication technologists and on the accuracy desired.

As mentioned in Sec. 5.2.2, in some technologies direct contact on top of the capacitor's top plate is allowed. A unit capacitor in such technologies is laid out as in Fig. 6.16*a*. The top connection extends in both directions, to allow wiring to the other unit capacitors or to other devices, which may be on either side. Once one unit capacitor needs to

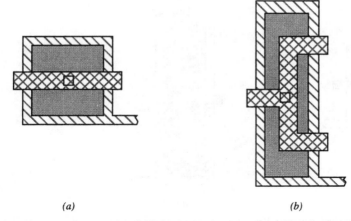

(a)   *(b)*

Figure 6.16 Layout of capacitors in a technology that allows contact over the top plate. In this example, the total parasitic capacitance of the necks is proportional to the main capacitance.

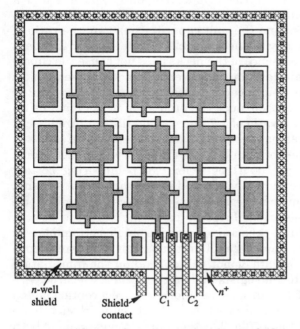

n-well
shield   Shield C_1 C_2
                contact

Figure 6.17 A layout of two capacitors with ratio $C_2/C_1 =$ ½, incorporating several of the techniques discussed in the text (adapted from Ref. 23).

be laid out in this way, all unit capacitors related to it should be laid out in this way, too. A nonunit capacitor equal to 1.5 units, following the rationale presented above, is shown in Fig. 6.16b.

Example 6.7   Lay out two capacitors C_1 and C_2 such that $C_2/C_1 = 3.5$.

solution   A capacitor layout that combines various techniques mentioned above is shown in Fig. 6.17 (adapted from Ref. 23). The layout provides an accurate capacitor ratio of $\frac{7}{2} = 3.5$. Narrow dummy capacitors are used around the periphery. Note the use of the connecting necks in the top plates; each unit capacitor has four such necks, even at the edges of the array where no connections need to be made. Thus the extra capacitance contributed by these necks is the same for each unit capacitor. The metal connections to C_1 and C_2 add some parasitic capacitance to them; if desired, these parasitic capacitances can be made to be in the same ratio as C_1 and C_2 by choosing the length of the metal connections accordingly. A bottom shield is used for interference reduction (Secs. 5.2.2 and 6.5).

The capacitor ratio matching accuracy that can be achieved depends on the fabrication technology, the size of the unit capacitors, the ratio desired, and the care with which techniques such as the ones described above are applied. Particular techniques may work better with certain technologies; for example, as already mentioned, the approach in Fig. 5.6 may work or may make matters worse, depending on the technology. If in doubt, it is best to use only 90° corners. Ratio accuracies of better than 0.1 percent are possible between two large arrays of unit capacitors, interleaved by using a common-centroid geometry layout. However, the matching between two unit capacitors, or between a unit capacitor and an array of such capacitors, is typically about 0.5 percent. The references provide additional material on capacitor ratio accuracy and layout techniques [5, 6, 12, 14–19].

6.4.2   Resistor layout

Resistor matching is achieved along similar lines as above. Again, identical unit resistors including identical contacts, etc., are used; these are connected in series or in parallel. Interleaving the unit resistors is important in order to reduce the effect of not only process gradients but thermal gradients as well; the latter can be caused by other devices on the chip and can significantly affect matching (Sec. 6.4.4). Dummy resistors can be used on either side of a resistor array, for the same reason dummy capacitors are used in capacitor arrays (see the previous section). In some instances, matching as good as that for capacitors has been reported. Of course, for good matching minimum dimensions are avoided, to reduce the effect of edge variations. It is not uncommon to use resistor widths of 10 or 20 μm, even in fabrication processes that theoretically allow for 2-μm resistors. Spacing between resistors, including spacing between main resistors and dummy ones, should also be somewhat greater than minimum (to avoid processing imperfection that can result from the close proximity of structures) and should be the same everywhere.

Example 6.8   A layout of two well resistors R_1 and R_2 with $R_2/R_1 = \frac{3}{2}$ is shown in Fig. 6.18. If R is the resistance of the main body of one unit section and R_C is the resistance of each end club head (including the effective resistance of the pair of contacts), then we have $R_2 = 3R + 6R_C$ and $R_1 = 2R + 4R_C$. Thus $R_2/R_1 = \frac{3}{2}$ independent of the values of R and R_C (which are not known accurately anyway). Note that, taking advantage of required well spacing, the connections of the various sections are made in such a way that current traverses all sections in the same direction; this precaution has been claimed to improve matching in some cases. However, it often does not make a significant difference and is ignored.

Unfortunately, in some cases even the approach illustrated in Example 6.8 is not adequate. The problem can be that even though the number of contacts has been properly considered, their resistance can vary *widely* from contact to contact (e.g., by a factor of 5 or more in some processes). In that case, the only way to keep the mismatch low is to ensure that the contact resistance is much smaller than the resistance of the resistor's main body. If this is not the case, one may have to use multiple contacts in an attempt to keep the effective contact resistance, and thus its absolute variation, low. The resistance of vias should be similarly considered.

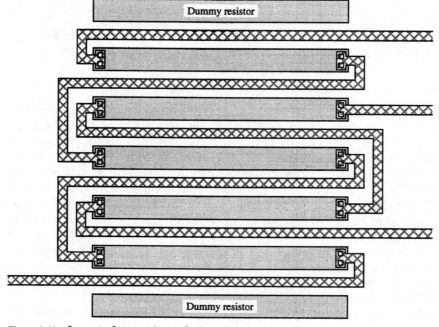

Figure 6.18   Layout of two resistors for a resistance ratio of $\frac{3}{2}$.

Another factor that can influence matching is mechanical stress caused by packaging. This effect is less pronounced in certain crystallographic directions along the wafer's surface; if sufficient information is provided, resistors can be laid out along such directions in order to reduce the magnitude of stress-related effects.

6.4.3 Transistor layout

Layout for transistor matching follows rules similar to those described for resistors and capacitors. Keep in mind that the gate length L and gate width W are not accurately known (Sec. 6.2); l_r, l_{lat}, and ΔW in Eqs. (6.1) to (6.3) are never known accurately, and they exhibit a random variation from device to device. Thus, for two transistors to be accurately matched, one should make L and W larger than the minimum allowed values (which can be, respectively, 1 and 2 μm for a "1-μm" process), to keep the effect of this randomness small. And, obviously, it is *not* enough for the two transistors to have the same W/L value; *they should have the same W and the same L.* This will also ensure that other effects, such as short- or narrow-channel effects, will be the same in both. Of course, for two identical devices to produce identical currents, they should be biased identically, too.

Let us now look at the mismatch effects in more detail [15, 17, 20–26]. Mismatches can occur due to local random effects (such as nonflatness of poly due to granularity) and to global effects such as oxide thickness gradients over the chip. The local random effects tend to average out in large devices; the global effects tend to be negligible for devices placed close together. Thus, for good matching one should use transistors of nonminimum dimensions, placed close together (although not *too* close, in order to avoid proximity effects during processing).

Current matching can be inferred from the matching of parameters V_T and k in the saturation current equation (Table 3.1). It has been found that for devices with channel lengths not too close to the minimum allowed by the technology, the contributions of local random effects to the mismatches ΔV_T and $\Delta k/k$ are inversely proportional to $\sqrt{WL}$, where WL is the channel area [22]. If, for example, in a 1-μm process two closely spaced 2-μm by 2-μm transistors have $\Delta V_T = 10$ mV and $\Delta k/k = 1$ percent, then by increasing their size to 20 μm by 20 μm one may roughly expect $\Delta V_T = 1$ mV and $\Delta k/k = 0.1$ percent. These larger devices, biased identically and with V_{GS} 1 V above threshold, will provide currents matched within about 0.3 percent; in weak inversion, the mismatch will be about 3 percent. These numbers can be calculated from the assumed ΔV_T and $\Delta k/k$, by using the equations in Table 3.1. In some processes, increasing the distance between the small transistors to a couple of hundred micrometers may not affect the mismatches much, since global effects are masked by random local effects. For the large devices, though, those random effects are small, and increasing the distance between the devices makes the global effects felt. Thus, separating the 20-µm by 20-µm transistors by a few hundred micrometers could double their ΔV_T and $\Delta k/k$ mismatches [22].

Another effect to be considered is the orientation of the transistors. Because crystallographic properties and process fabrication details are not identical in different directions along the surface, in devices to be matched the currents should flow parallel to each other (and ideally in the same direction). Mismatches ΔV_T and $\Delta k/k$ can increase several times if the devices are laid out perpendicular to each other [22].

The layout of transistors with accurate current ratios (under identical bias conditions) follows the ideas discussed earlier about capacitors and resistors with accurate ratios. Thus, for example, to make a transistor with "precisely" twice the current of a smaller one, one should actually make two transistors identical to the latter and connect them in parallel. One way to do this is shown in Fig. 6.19. Note that the shapes of all unit devices are identical. The right-hand combination offers two paths to the current, each of which is identical to the left-hand path not only in terms of the intrinsic part, but also as far as the extrinsic parts are concerned, including contacts. The orientation of the devices is also identical; this improves matching, as already mentioned.

The devices in Fig. 6.19 can be interleaved in an attempt to further improve matching, by using the center device by itself and connecting the leftmost and rightmost devices in parallel; this, however, would complicate the layout (Prob. 6.13). Such interleaving will be worth doing if the outer devices are rather far apart (e.g., by more than 100 µm), in

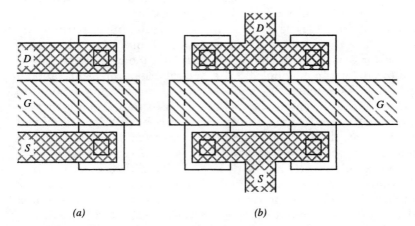

(a)   (b)

Figure 6.19 Simple layout of two MOS transistors intended to provide a drain current ratio of 2 under identical voltage bias. First-order gradients in the horizontal dimension do not cancel, but are not expected to be a dominant source of mismatch if the devices are not large, and if thermal gradients are absent.

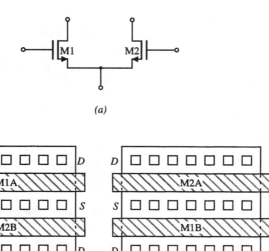

(a)

(b)

Figure 6.20 (*a*) A differential pair; (*b*) common-centroid layout for attempting to make M1 and M2 identical.

which case the fabrication processing conditions for them may differ significantly, or if thermal gradients are expected (discussion follows).

Common-centroid geometries are sometimes used if two devices M1 and M2 must be as identical as possible, such as in the input stage of a differential amplifier. This makes sense if the transistors are large, in which case global effects are felt. This is illustrated in Fig. 6.20. The devices in Fig. 6.20*a* can be laid out as shown in Fig. 6.20*b*. Four identical subdevices are used, with shared sources. M1A and M1B are connected in parallel to form M1, and so are M2A and M2B to form M2. This "cross-coupled" connection reduces the effect of process parameter and thermal gradients in any direction. The connections between drains and between gates are not shown; they are left for a problem (Prob. 6.15). Note that using the outer strips as drains increases the drain parasitic capacitance (Sec. 6.3).

Other effects to be considered include the variability of contact resistance, already discussed in relation to resistor matching in Sec. 6.4.2. Mechanical stress, also discussed for resistors in Sec. 6.4.2, can affect transistor matching, too, and so can unequal heating by nearby structures (other transistors, resistors, or even interconnect wiring, which is a heat conductor). It has also been observed that the mere proximity of transistors to other structures (even as far away as 20 μm or more) can affect transistor matching, even if such structures do not cause heating. This can be traced to the nonuniformity of photolithographic processing, caused by proximity effects [26]. Also, mechanical stress caused by the proximity of wide and long interconnect wiring (even at distances as large as 50 μm) can affect matching [26]. For these reasons, transistors which are supposed to be highly matched are sometimes kept well away from other structures and unrelated interconnect wiring.

6.4.4 A collection of rules for good matching

The above discussion of component-matching techniques is not meant to be exhaustive. Particular circumstances may require the invention of new techniques, and the designer's ingenuity can make a real difference. As already mentioned, the central idea is to *make everything identical so that all components to be matched are affected by the same factors in the same way.* This implies that, in aiming at matched components, the following list of considerations should be kept in mind [24, 12, 26].

1. *The devices to be matched should be of the same type.* One cannot hope to match nMOS transistors to pMOS transistors, or oxide capacitors to junction capacitors.

2. *They should have the same shape and the same size.* As already mentioned, for example, it is not enough that transistors to be matched have the same W/L ratios; they should have the same W and the same L. Even the connections to the devices (length and shape) should be identical in certain cases, if at all possible. Special geometries, such as those in Figs. 5.6, 6.5b, and 6.5c, should be avoided unless there is satisfactory information about their matching properties in the fabrication process to be used.

3. *They should not be of minimum size.* This means they should be designed with dimensions larger than the minimum resolution possible in the fabrication process, to reduce the influence of local random effects.

4. *They should be located close to each other.* (Although they should not be as close as the minimum feature size specified for the technology at hand.) This will ensure that fabrication conditions are nearly identical. For transistors to be placed inside wells, a common well should be used if their bodies are electrically common and there is no possibility of interference.

5. *They should be laid out using common-centroid geometries where warranted.* This rule is especially important for large element arrays, in which points at opposite extremes are distant (100 μm or more).

6. *They should have the same orientation.* For example, the current flows of transistors to be matched should be parallel (and ideally in the same direction). This will avoid problems with nonisotropic fabrication processing and even nonisotropy of the silicon substrate.

7. *The effect of the environment should be carefully considered.* Dummy components can be used at the ends of device arrays, at dis-

tances equal to those between the main devices themselves. Examples have been shown in Figs. 6.13c, 6.17, and 6.18. Interconnect wiring should not pass above transistors that are supposed to be highly matched (or, if it has to pass above them, it should at least be made of a high-level metal layer). In critical cases, matched transistor arrays may have to be laid out far away from other structures and from wide and long interconnect wiring, unless it is known that proximity to these does not influence matching in the fabrication process to be used.

8. *They should be at the same temperature.* If there is significant power dissipation by other devices on the chip, the devices to be matched should be placed as far as possible from those, and should be laid out symmetrically with respect to them so that they are heated equally. This is illustrated in Fig. 6.21. Also keep in mind that although a metal wire may not generate much heat, it may conduct heat generated elsewhere.

9. *The effect of the contacts and connections to the devices should be carefully considered.* The same number of contacts should be used in each device and in the same place. If vias must be used, ideally the same number of vias should be used in the wiring of each device. Multiple contacts should be used to combat the effect of contact resistance variability, wherever it is expected that the latter may occur. The "wires" connected to the devices should be the same type, and the parasitic resistance (and capacitance, for *ac* work) of these wires should ideally also be matched.

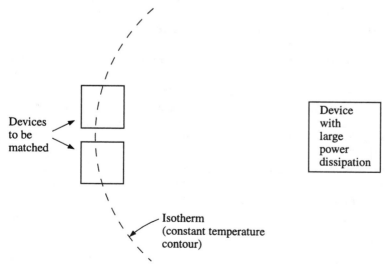

Figure 6.21 Placing devices symmetrically with respect to a heat-producing device.

10. *They should be biased identically.** In particular, to achieve identical currents, it is not enough that MOSFETs to be matched have the same gate-source voltage. They should have the same substrate-source and drain-source voltages as well. Identical biasing will also help in the case of passive devices, especially if their voltage coefficient is large.

Although we have discussed how to achieve matching, we have not considered *why* matching is desirable or which devices should be matched in a circuit. The reader will find many such considerations in circuit design texts [5–10].

6.5   Layout for Interference Reduction

VLSI chips consist of several subsystems which must be kept from interfering with each other. Unfortunately, the close packing of so many components on a chip, and their sharing of a common substrate and common connections to the external world, makes interference very likely, unless special precautions are taken. This interference is often referred to as "noise," without implying randomness. The interference problem is especially serious when analog and digital circuits are mixed on the same chip, which is increasingly the case in modern VLSI. Analog circuits contain extremely sensitive circuits, e.g., op amps and comparators, which can take a few microvolts of signal or *interference* at their inputs and convert them to several volts at their outputs. Digital circuits, on the other hand, operate with rapidly switching waveforms with swings of several volts. It stands to reason that if even a minute fraction of such swings is coupled to sensitive analog nodes, the results can be disastrous for the affected analog circuit and thus for the performance of the entire VLSI system.

Parasitic coupling of interference can occur through many paths, for example, the inductance and resistance of power supply or ground lines common to an interfering circuit and a circuit sensitive to interference, or the parasitic capacitances between two such circuits. Another big culprit is usually the common substrate. Currents can be caused in it by a variety of sources, including parasitic capacitances of circuit elements, power supply buses, bonding pads, and electrostatic discharge (ESD) protection circuits; in the presence of rapidly changing voltage waveforms, such capacitive currents can be large. Currents can also be created in the substrate from "hot" carriers escaping high-field regions near transistor drain junctions, or from carriers that escape from the channel into the substrate when a digital or analog

---

* This last rule is not directly related to layout, but it is included here for completeness.

switch is turned off. When substrate currents flow through the substrate resistance, they cause voltage drops which can influence other devices on the common substrate. In some cases, spikes of several volts can be present on ground and power supply lines and on the substrate [27–29]. Needless to say, *trying to make analog circuits coexist with digital circuits on the same chip under such adverse conditions can be a nightmare, unless the designer takes all precautions imaginable.* This is especially true in view of the fact that interference is difficult to predict or simulate.

Several strategies have been developed for attacking the above problems. Some are intimately related to the circuit design process. For example, the timing of the circuits is chosen so that comparators are not comparing, and sample-and-hold circuits are not sampling, at instants when large digital drivers switch. Also, fully differential circuits are used in which the usable signal is the difference between the signals on two lines; interference that is common to both lines cancels out in the difference [5–10]. The use of special logic circuit families which produce less interference than common CMOS logic has also been proposed [30]. Success with eliminating interference, though, also resides to a very large extent with careful layout techniques [5, 12, 23, 28, 29, 31–38]. Several such techniques are described below.

1. *Avoid common power supply or ground buses* between circuits that can interfere with each other. This is *extremely important* to keep in mind [32–35]. Examples of bad and good practice in this respect are shown in Fig. 6.22. In Fig. 6.22*a*, a common power supply line is used for both analog and digital circuits. The digital part contains digital bus drivers or output drivers, which must charge large capacitances over large voltage swings in a short time. The resulting currents, assuming the capacitances are linear, are given by $i = C \, dv/dt$ [39] and contain large spikes as a result. When these spikes pass through the power supply line and package inductance* and resistance as shown, they create across them significant voltage drops (e.g., a few tenths of a volt or even more), especially if they are varying fast, since the voltage across an inductance L is $L \, di/dt$ [39]. These voltage drops appear in series with the power supply voltage. Thus the analog circuits are not fed by a clean dc voltage, but rather by one with "noise" added to it. The noise then finds its way to the analog signal paths because of poor power supply rejection in the analog circuits, parasitic coupling, etc. At times the problem can be made worse by the fact that bonding wire and pin inductance, combined with the parasitic capacitance of power supply buses, pads, and ESD protection circuitry, can form a resonant LC

---

* Typically several nanohenrys.

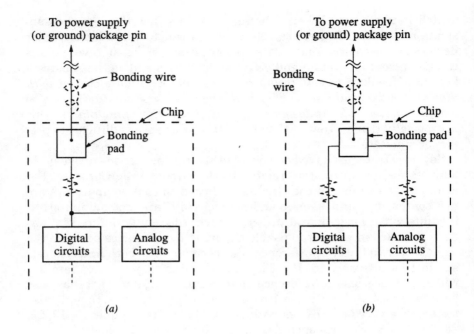

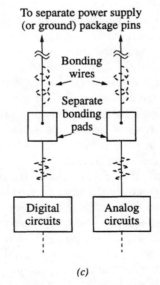

Figure 6.22 Interference from digital to analog circuits on the same chip. (*a*) Bad connection, with common on-chip wiring and bonding wire; (*b*) improved connection, eliminating the common on-chip wire; (*c*) best connection, with separate on-chip wires, bonding wires, and pins.

circuit [37]. If the resonant frequency of this circuit happens to be equal to the frequency of interference (e.g., the digital clock frequency or one of its harmonics), then the impedance of a parallel resonant circuit can become very large [39], and thus the above problems can worsen. Simulations of the complete circuit, including the pads, protection circuitry, and a model for the package, should be used to check for such troublesome possibilities.

To reduce the magnitude of the above problems, the strategy in Fig. 6.22b can be used. Now there is no coupling due to the on-chip wiring, but still the bonding wire is common and much of the interference remains. A better approach is shown in Fig. 6.22c, where two separate pads are used, connected to two separate package leads (or to the same lead, if two separate ones are not available). The leads are then externally tied together and connected to the power supply. Now the only shared impedance is that due to the external connection and power supply output.* If that still causes a problem, bypass capacitors may have to be connected externally to the pins. Internal bypass capacitors are used sometimes. Of course, every time a bypass capacitor is added or a bonding wire inductance is changed (see item 2 below), one should make sure that undesired resonances do not occur.

Similar problems will occur with ground buses, and they can be attacked by using the same techniques. Thus, one should use separate *digital ground* and *analog ground* lines. In general, *to properly lay out a VLSI chip, one has to stop thinking of power supply lines and grounds as perfect conductors.*

Common lines can cause interference also from one analog circuit to another. For example, analog switches charging capacitors can cause large currents and thus large voltage drops on ground lines. If the input terminal of an op amp is connected to such ground lines, it will pick up the interference which will subsequently be amplified. Thus, separate ground lines must be used in this case. In some cases, it is possible to group elements that can use a common ground or power supply line. Thus, e.g., it may be possible to group all op amps in one area, all switches in another, all capacitors in yet another, etc., and feed each group with separate lines. This is often done with switched-capacitor circuits [5, 12]. It is also advisable that the drivers of analog switches be fed from "clean" power supply lines; otherwise, power supply "noise" can be coupled to the analog circuits through the switches.

2. *Assign bonding pads wisely.* When a chip is attached to a package, some bonding pads are close to their corresponding package pins

---

* This solution can cause an unexpected problem. When the power supply is turned on, differences in the two paths may result in some parts on the chip "seeing" the supply voltage before others do, and this can activate latch-up [12]. Adequate protection against latch-up should thus be used (Sec. 5.2.9).

whereas others are not so close. The package bonding diagram should be anticipated, and for connections in which the bonding wire inductance is expected to cause problems (see item 1 above), one should use pads to which short bonding wires are to be connected. To reduce the inductance associated with bonding wires for power supply and ground lines, several bonding wires and pads can be connected in parallel.

3. *Make power supply lines and grounds wide,* to ensure a low-resistance path. This, for example, might be especially helpful for current-carrying ground wires to which op amp inputs are tied. In certain troublesome cases, lines as wide as 100 µm are used. The increased parasitic capacitance of such lines to the substrate, though, must be carefully considered, since it can cause noise coupling to or from the substrate (see next item).

4. *Consider substrate coupling carefully, and guard against it.* Varying voltages on device junctions, wiring, and pads can be coupled capacitively to the substrate. As an example, consider the situation in Fig. 6.23a. The junction on the left is connected to a widely varying voltage (this junction can be, for example, the drain of a transistor belonging to a digital circuit); in the figure, a voltage source is used to represent this voltage. Through the junction capacitance, a varying voltage V is then caused to exist below the junction, as shown in the figure. Since the substrate is not an insulator, part of the voltage variation appears below the transistor on the right and is coupled to it through parasitic capacitances and through the body effect (because the device's V_{SB} is varying—see Secs. 2.4.2 and 3.2.2). This interference will be passed to the circuit of which the transistor is part. If the device on the right were a resistor or a capacitor, interference would similarly affect it through parasitic capacitance and (in the case of well and diffusion resistors) through a nonzero voltage coefficient.

To reduce the above problem, consider placing a grounded substrate contact next to the interfering device on the left, as shown in Fig. 6.23b. Although the substrate resistance is distributed in three dimensions, we will use a simplified picture with lumped resistance in order to offer a qualitative explanation. If the contact is close to the device, the substrate resistance R_1 between them will be small, and thus the voltage V_1 will be greatly reduced in comparison to V in Fig. 6.23a (ideally, if R_1 were a short circuit, V_1 would be zero). Thus the interference being "transmitted" toward the right is reduced. If now we place a grounded substrate contact next to the interference-receiving device on the right, as shown in Fig. 6.23b, the interference will be reduced even further since R and R_2 in the figure form a voltage divider, with $V_2 = V_1 R_2/(R_2 + R)$.

To increase the effectiveness of the above approaches, one can use two strategies: (1) increase R and (2) decrease R_1 and/or R_2. To increase

R, one can increase the separation between the interference transmitter and receiver. In common fabrication processes with a uniform lightly doped substrate, this works well, and separations of up to a few hundred micrometers can be used to advantage. However, in processes which use a thin epitaxial layer above a heavily doped substrate (in order to combat latch-up—see Sec. 5.2.9), this approach is not very effective, as can be seen from Fig. 6.23c: here the p^+ substrate, with its low resistance, is effectively in parallel with the resistance R of the epi-

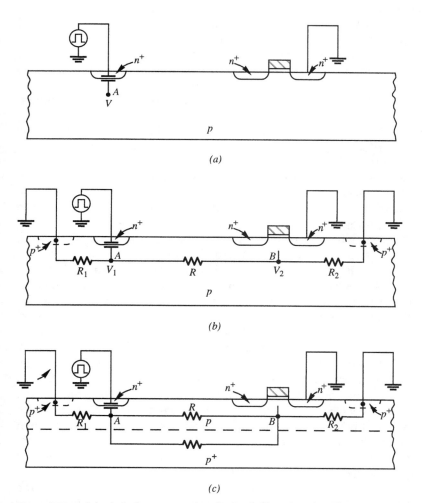

Figure 6.23 (a) An interference-causing device (left) and an interference-sensitive device (right) on a common lightly doped substrate; (b) use of substrate contacts for substrate coupling reduction; (c) the above devices implemented in a fabrication process which uses a thin epitaxial layer on top of a heavily doped substrate.

taxial layer, and nodes A and B are practically short-circuited. In this case, separations of more than a few times the epitaxial layer thickness do little to reduce interference [34, 37]. The thickness of epitaxial layers can be as small as 5 to 10 μm in some processes. In such cases, the only way to prevent the transmitter from "talking" to the receiver is to carefully ground the heavily doped substrate, thus maintaining nodes A and B at nearly zero potential. Such grounding can best be accomplished with a chipwide back contact at the bottom of the wafer. Not every technology, though, and not every package type allow for such contact. Also, it may not be possible to establish a good ground connection to the back of the wafer during water-level testing.

A very effective way to increase R (practically to infinity) is to use an insulating substrate, such as provided in a silicon-on-Insulator (SOI) process. Such processes are far from standard, though. Also, at very high frequencies much of the insulation is lost, due to parasitic capacitances.

The second interference reduction strategy, i.e., reducing R_1 and/or R_2, can be effectively implemented by forming the substrate contact as a guard ring which surrounds the devices or circuits that transmit and/or receive interference. This is shown by the inner ring in Fig. 6.24. The ring should be close to the circuit it guards. It should be strapped with metal throughout its length and should be connected to the metal through multiple contact windows. An example of a device that can greatly benefit from a guard ring around it is a well resistor. Substrate potential variations can greatly interfere with proper operation of this resistor, not only through parasitic capacitance but also through the fact that they can modulate the width of the depletion layer between the well and the substrate and thus can modulate the area of the cross section through which the current passes; this directly modulates the resistance value. To keep the substrate potential in the vicinity of the well resistor constant, then, it should be surrounded by a grounded guard ring. If multistrip well resistors are used, it may be necessary to surround each strip individually by a guard ring [12].

Increasing the width of a guard ring decreases R_1 or R_2 even further. However, here care is required. If the guard rings around interference transmitters and receivers are made so wide that the rings approach each other, this will effectively reduce the resistance between them and can then *increase* interference rather than reduce it; this would correspond to decreasing R in Fig. 6.23b.

Guard rings around entire interference-producing circuits (e.g., around the entire digital section of a chip) can be an important help. Sometimes interference-producing circuits can be laid out in a way that results in self-shielding; for example, a digital output buffer can be laid out with the well of the nMOS device nearly surrounding the pMOS device [12].

Guard rings around sensitive devices must, of course, be connected to "quiet" grounds in order to be effective. Such grounds are not easy to come by, due to packaging inductance (see item 1 above). In critical cases, one may have to use separate connections to guard rings and bring them out to dedicated package pins. The biasing of guard rings around interference transmitters requires extra care. To make a guard ring bias quiet, one may attempt to connect it to an analog ground. In that case, however, the guard ring may collect interference from the noisy devices it surrounds and may feed that interference to other circuits through this line. This problem is avoided if a digital ground is used; however, the bias potential of the shield will not be very quiet. Which solution is adopted will depend on a careful assessment of the situation at hand.

A second type of guard ring, consisting of an n-well, is sometimes used; see the outer ring in Fig. 6.24. This ring can be grounded, or it can be connected to a clean positive potential. Its main purpose is to interrupt the flow of surface substrate currents, which can be large because the substrate doping is higher near the surface due to the presence of the field implant (Sec. 5.2.1). Well rings may also act as collectors for stray carriers in the substrate. However, if a p^+ ring is used as suggested above, the effect of adding an n-well ring is sometimes limited [34, 35].

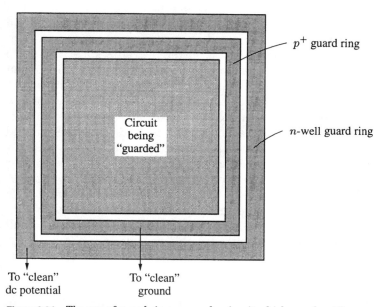

Figure 6.24   The use of guard rings around a circuit which may be either an interference transmitter or an interference receiver.

To reduce the possibility of latch-up (Sec. 5.2.9), it is common to make contacts at regular intervals between the $-V_{SS}$ lines (for an n-well process) and the p substrate [40, 41]. While in general this is good practice, it can create interference-coupling paths between a $-V_{SS}$ line and the substrate, or from one $-V_{SS}$ line to another (e.g., from one feeding digital circuits to one feeding analog circuits) through the substrate resistance. If such problems are suspected, one can use a $-V_{SS}$ line without substrate contacts to feed the sensitive circuits and another low-resistance line with contacts to the substrate, connected only to a $-V_{SS}$ pad (where the voltage is cleaner). This would be especially advisable to do when the $-V_{SS}$ line feeds devices which can be "transmitters" of interference. In the case of devices which can be "receivers" of interference, the situation calls for careful judgment: Since separating the sources from the substrate can allow the existence of a potential difference between the two, substrate interference can affect the devices by modulating their threshold voltage through the body effect.

It is evident that in devising ways to reduce substrate coupling, one can benefit from a good simulation tool. Since the effects described are three-dimensional, accurate simulation is very difficult and can be painfully slow. One can, however, use certain simple techniques to approximately simulate substrate coupling by using common circuit simulators [34–35].

5. *Shield all sensitive circuits, devices, and interconnection lines.* Use the bottom plate shields for resistors and capacitors discussed in Secs. 5.2.2 and 5.2.3. Use shields for sensitive lines. For example, a metal line can be (partially) shielded by two other lines, one on each side, and by a poly line below it (the latter prevents substrate noise from being coupled to the metal line). In some cases, top shields can also be used; for example, in double-metal processes a sensitive poly line can be shielded from a second-level metal line over it by using a first-level metal shield between the two. Transistors sensitive to substrate noise should ideally be placed inside wells (and thus they should be pMOS transistors in an n-well process). Wells of sensitive devices should be tied to the devices' sources or to a "clean" potential, rather than a noisy power supply line. An example is the circuit of Fig. 6.7. Separate wells should be used for transistors that can interfere with each other. For devices that sit directly on the substrate and thus cannot be shielded from below (such as n-channel transistors and n-well resistors on a p substrate), the only possible protection is a guard ring. This type of shield has already been discussed in item 4 above; as explained there, it can be key to interference reduction.

6. *Shield all interference-producing circuits,* using techniques as in item 5 above. Thus, for example, even if the interference picked up from the substrate by the bottom plate of a capacitor does not cause

concern, it is important to shield the bottom plate (Figs. 5.3c and e) so that the capacitor cannot *induce* interference into the substrate which could then affect other circuits. Analog switches with their gates connected to clock lines are "semidigital" circuits, and they must be treated carefully; e.g., their wells should be separate from those used for op amps or capacitor shields.

7. *Avoid proximities of circuits, devices, or interconnection lines that can interfere with each other.* This has already been discussed above in relation to substrate coupling; but it is, in fact, a general strategy to reduce other types of coupling as well, e.g., electrostatic coupling between metal wiring or between the tops of capacitor plates. Sensitive interconnection lines should not run next to, or cross, lines with signals that can cause interference. Sometimes distances as large as 100 µm must be maintained.

8. *Make the lines connected to sensitive points as short as possible,* to prevent them from acting as receivers of interference. For example, if a line connected to an op amp input is long, it can pick up interference from nearby noisy lines through electrostatic or magnetic coupling.

9. *Make the lines connected to interference-producing circuits as short as possible,* to prevent them from acting as transmitters of interference. This is particularly important for the power supply and output leads of digital circuits driving large capacitances.

10. *Consider the connection of devices with nonsymmetric parasitics carefully.* For example, consider an integrated capacitor in which the bottom-plate parasitic (Fig. 5.4) can couple noise from the substrate (as in Fig. 5.3b or d). Assume that the main capacitor must be connected from the negative input to the output of an op amp. If the capacitor is connected as in Fig. 6.25a, "noise" on the substrate (which is connected to $-V_{SS}$) will be coupled to the amplifier input through the parasitic capacitor and can be subsequently amplified. If the connection is reversed, as shown in Fig. 6.25b, the interference coupling will be to the (normally low-impedance) output node, so the problem will be less severe. (Now the top plate parasitic—not shown—will be at the input, but that parasitic is normally much smaller.) However, the voltage at the output of the op amp can be coupled to the substrate and through it to other circuits on the chip. A still better approach is to use a capacitor with a shielded bottom plate, like the ones in Figs. 5.3c and e, and to ground the shield as shown in Fig. 6.25c, provided, of course, that the ground used is quiet (see item 1 above).

11. *Lay out fully differential circuits as symmetrically as possible.* As mentioned earlier, the success of fully differential topologies lies in taking the difference between the signals on two lines; if interference common to both lines exists, it cancels out in the difference. Thus, the layout should be such as to ensure that the interference is, indeed,

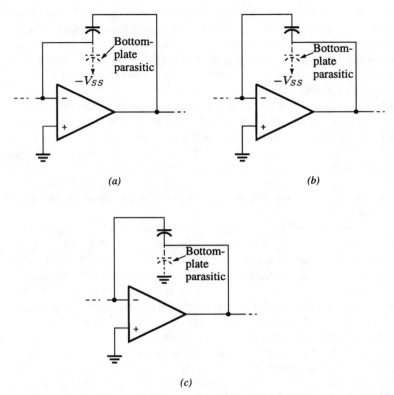

Figure 6.25 Connecting a capacitor in the feedback path on an operational amplifier. (*a*) Wrong connection, in which the bottom plate parasitic can result in interference from the substrate; (*b*) better connection; (*c*) best connection, using a bottom plate shield.

common. This makes necessary the matching of not only the two signal paths, but also the *parasitics* associated with these paths. While always a completely symmetric layout would be the obvious choice, this may not be possible due to layout constraints.

12. *Consider carefully the "floor plan" of the entire chip.* This will in general require separating the analog and digital parts of the chip; see, e.g., the chip photographs in Figs. 1.1 and 1.3. One must also consider "sorting" the various sections within each part wherever possible. Consider, e.g., a mixed analog-digital circuit which must accept minute analog signals at its input and must drive digital signals into heavy loads at its input. A sensible way to lay out such a circuit is shown in Fig. 6.26 [38]. Here the worst interference transmitter is the section containing the digital output drivers, and the most sensitive interference receiver is the analog input section. Thus, these sections have been placed as far as possible from each other, at opposite ends of a chip diagonal.

Medium-amplitude analog circuits are less sensitive, so they are allowed to be closer to the digital circuits; however, they are not placed as close to them as high-amplitude analog circuits, which are the least sensitive analog blocks. Of the digital logic circuits, low-speed ones produce less interference than high-speed ones; the former are thus placed closer to the analog section. In the floor plan, guard rings are placed around entire sections, for reasons already discussed. In addition, guard rings may be used around critical areas within each section.

From the above discussion it is clear that interference reduction requires the consideration of a multiplicity of problems, some of which are interdependent. The relative magnitude of these problems, and the suitability of the various solutions proposed, will depend on the details of the circuits being implemented. Thus one cannot be more specific at this point; the best that can be hoped for is that the reader has become aware of the various issues involved. By following the above guidelines and applying extra judgment to the case in hand, a lot of potentially troublesome interference can be avoided. Further layout considera-

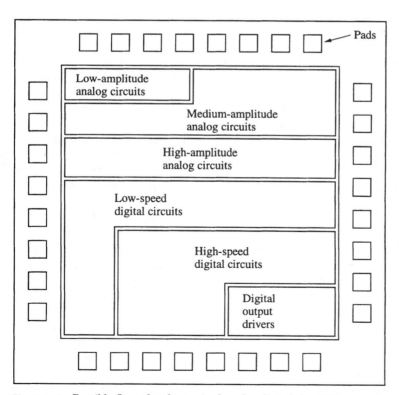

Figure 6.26 Possible floor plan for a mixed analog-digital chip [38].

tions for specific types of circuits are discussed in books where such circuits are covered [5, 6, 12].

6.6 Integrated-Circuit Design

The design of an integrated circuit starts with certain objectives to be met (specifications, or "specs"). To meet these objectives, the circuit designer decides on a system architecture and interconnects building blocks to implement it. The designer may have to do a "full-custom" design, i.e., may have to design such blocks from scratch. Or perhaps she or he can use predesigned, existing *standard cells* stored in computer libraries. Such libraries usually contain often used blocks (counters, operational amplifiers, comparators, etc.) and the library user may be provided only with the input-output characteristics and/or macromodels for these blocks. Standard cell libraries make VLSI accessible to system designers who may not have the knowledge to design the cells themselves.

In general, from-scratch, full-custom design done by a competent designer can be expected to produce integrated systems with higher performance. This is because in full-custom design each building block can be individually optimized so as to best serve the needs of the system. In contrast, in standard-cell design the designers may have to deal with generic cells designed for use in a variety of situations, thus representing compromises. Sometimes, however, the extra benefits that can be derived from designing everything from scratch (e.g., smaller chip size) may not be worth the extra design time. This may be the case, e.g., if quick availability of the chip is of extreme importance; it can also be the case if a chip is to be produced in limited quantities, in which case the design effort may be a significant part of the total production cost per chip. The opposite may be true in large-volume chips, where design time costs are shared among millions of identical chips.

A compromise between the full-custom and standard-cell approaches is to use *parameterized cells*. These are cells with largely predetermined designs, in which certain parameters can be altered to attain a desired performance (e.g., increased op amp speed at the cost of extra power dissipation). Still, though, state-of-the-art performance usually requires that circuits be designed at the device level, at least for part of the chip.

In addition to the above approaches, there is the *array* approach. In this, fixed, commonly used devices or building blocks are already available on chips, which exist on prefabricated wafers with only the metal interconnection layer missing. Information on what is available on such arrays is provided by the manufacturer to prospective users. The latter specifies to the manufacturer a wiring pattern for connecting appropriate blocks for a given application. The manufacturer then performs the final step in the fabrication process, i.e., metallization, according to the specified wiring pattern. The advantages of this approach are the low cost (since a large number of identical arrays are fabricated for many different users) and the implementation speed, both in the design and in the postdesign manufacturing. However, the design flexibility is necessarily limited and so may be the resulting performance. Also, for a particular application, not all the available blocks or interconnection space on the chip is likely to be used, which means wasted chip area. The array approach has been used mostly for digital circuits so far.

Circuit design may proceed according to the following scheme. After system design and system simulation, the specifications of the individual circuits, which comprise the system building blocks, are defined, and each individual circuit is considered separately. Following initial conception of a circuit and approximate hand analysis, a circuit simulator is used to predict its performance, using perhaps approximate estimates of device parasitics. If the performance predicted is not satisfactory, design changes are made and the circuit is simulated again. Such iterations may have to be repeated several times, until the performance objectives are met. By using a layout editor, the circuit is laid out in detail, by specifying the geometric shape and location of the various layers (sources and drains, gates, wells, interconnections, etc.). A design rule checker is used to check whether any layout rules (Sec. 6.3) have been violated. Router programs can be used to help route the interconnections between circuit blocks. A circuit extractor program is used to reconstruct the circuit from the layout, including parasitic elements (capacitances to the substrate or between interconnection lines, etc.). Of course, such detailed information can be produced only at this stage, since the exact layout was not known during earlier design stages. The output of the circuit extractor program is fed to a circuit simulator, and the circuit is simulated again. If a problem is identified, the layout or even the circuit design itself is modified. In all simulations, it is of course important to include models for the pads, protection circuitry, and package. When the process converges and the layout-extracted circuit shows satisfactory performance, the circuit can be sent (in the form of a computer file containing the final layout information) for fabrication. Obviously, many boring and frustrating iterations can be eliminated if the design is done professionally from the beginning, using clever circuit techniques, careful layout, and interference reduction techniques.

For circuits that are meant for production, extensive simulations are needed that use both typical and worst-case parameters. This is to ensure that the design objectives are met over the required temperature and supply voltage ranges and in the presence of fabrication tolerances. Statistical simulations are also used. Again, good design practice can help ensure the success of such evaluations and avoid further iterations.

It is obvious that computer aids play a key role in IC design and that the many runs needed can be time-consuming. There is a strong trend to automate as much of the design process as possible. Programs already exist that will design common circuit blocks, starting from specifications and finishing with a layout. Even if the final design is not on a par with what a human designer would produce, at least it provides a starting point for a designer, who can now go ahead and improve it further, especially in the case of analog circuits; of course, detailed circuit design skills will be needed for this. The "touchup" performed by a human designer can involve changing certain device sizes to optimize a given performance parameter, placing farther apart devices that may interfere with each other, etc. Also some further compaction, resulting in smaller chip area, may be possible with the designer's intervention. This may or may not be worth the extra effort involved, depending on economic considerations (which, as already mentioned, include design time!).

Perhaps the extreme case of design automation is represented by *silicon compilers,* which are supposed to accept as input (in a high-level description language) information as to what a desired integrated system is supposed to do and to produce the final layout without human intervention. Such programs have met with success for certain digital systems. There remains, however, a lot of work to be done before this approach becomes viable for analog and mixed analog-digital circuits on a large scale [42].

In general, computer aids are an invaluable help, especially in relieving designers from routine work so that they can deal with more challenging circuit design tasks. There are plenty of challenges best suited for human designers, and this is not likely to change in the foreseeable future. It takes many years to automate existing design knowledge; in the meantime, new types of circuits are created by human designers, using perhaps the same principles learned on the old circuits and applied in a different way, or even using newly invented principles.

References

1. J. Y. Chen, *CMOS Devices and Technology for VLSI,* Prentice-Hall, Englewood Cliffs, NJ, 1990.
2. N. Weste and K. Eshraghian, *Principles of CMOS VLSI Design,* Addison-Wesley, Reading, MA, 1993.
3. L. A. Glasser and D. W. Dobberpuhl, *The Design and Analysis of VLSI Circuits,* Addison-Wesley, Reading, MA, 1993.
4. E. D. Fabricius, *Introduction to VLSI Design,* McGraw-Hill, New York, 1990.

5. R. Gregorian and G. C. Temes, *Analog MOS Integrated Circuits for Signal Processing,* Wiley, New York, 1986.

6. P. E. Allen and D. G. Holberg, *CMOS Analog Circuit Design,* Holt, Rinehart and Winston, New York, 1987.

7. M. R. Haskard and I. C. May, *Analog VLSI Design,* Prentice-Hall, Englewood Cliffs, NJ, 1988.

8. R. Unbenhauen and A. Cichocki, *MOS Switched Capacitor and Continous Time Integrated Circuits and Systems,* Springer-Verlag, New York, 1989.

9. P. R. Gray and R. G. Meyer, *Analysis and Design of Analog Integrated Circuits,* Wiley, New York, 1993.

10. K. R. Laker and W. M. Sansen, *Design of Analog Integrated Circuits,* McGraw-Hill, New York, 1994.

11. D. Senderowicz, "CMOS Operational Amplifiers", chap. 2 in J. E. Franca and Y. Tsividis, eds., *Design of Analog-Digital VLSI Circuits for Telecommunications and Signal Processing,* Prentice-Hall, Englewood Cliffs, NJ, 1994.

12. F. Maloberti, "Layout of Analog and Mixed Analog-Digital Circuits," chap. 11 in J. E. Franca and Y. Tsividis, eds., *Design of Analog Digital VLSI Circuits for Telecommunications,* Prentice-Hall, Englewood Cliffs, NJ, 1994.

13. C. A. Mead and L. Conway, *Introduction to VLSI Systems,* Addison-Wesley, Reading, MA, 1980.

14. J. McCreary, "Matching Properties, and Voltage and Temperature Dependence of MOS Capacitors," *IEEE Journal of Solid-State Circuits,* 16: 608–616, December 1981.

15. D. J. Allstot and W. C. Black, Jr., "Technological Design Considerations for Monolithic Switched-Capacitor Filtering Systems," *Proceedings of the IEEE,* 71(8): 967–986, August 1983.

16. J. B. Shyu, G. C. Temes, and K. Yao, "Random Errors in MOS Capacitors," *IEEE Journal of Solid-State Circuits,* 17: 1070–1076, December 1982.

17. J. B. Shyu, G. C. Temes, and F. Krummenacher, "Random Error Effects in Matched MOS Capacitors and Current Sources," *IEEE Journal of Solid-State Circuits,* 19: 948–955, December 1984.

18. R. Singh and A. B. Bhattacharyya, "Matching Properties of Linear MOS Capacitors," *Solid-State Electronics,* 32: 299–306, 1989.

19. M. J. McNutt, S. Le Marquis, and L. Dunkley, "Systematic Capacitance Modeling Errors and Corrective Layout Procedures," *IEEE Journal of Solid-State Circuits,* 29: 611–616, May 1994.

20. K. R. Lakshikumar, R. A. Hadaway, and M. A. Copeland, "Characterization and Modeling of Mismatching in MOS Transistors for Precision Analog Design," *IEEE Journal of Solid-State Circuits,* 21: 1057–1066, December 1986.

21. G. De Mey, "Stochastic Geometry Effects in MOS Transistors," *IEEE Journal of Solid-State Circuits,* 20: 865–870, August 1989.

22. M. J. M. Pelgrom, A. C. J. Duinmaijer, and A. P. G. Welbers, "Matching Properties of MOS Transistors," *IEEE Journal of Solid-State Circuits,* 24: 1433–1440, October 1989.

23. E. A. Vittoz, "Analog Layout Techniques," *Notes, Course on Practical Aspects in Analog IC Design,* Ecole Polytechnique Federal de Lausanne, Lausanne, Switzerland, July 1992.

24. E. A. Vittoz, "The Design of High-Performance Analog Circuits on Digital CMOS Chips," *IEEE Journal of Solid-State Circuits,* 20: 657–665, June 1985.

25. M. Steyaert, J. Bastos, R. Roovers, P. Kinget, W. Sansen, B. Graindourze, A. Pergoot, and E. Janssens, "Threshold Voltage Mismatch in Short-Channel MOS Transistors," *Electronics Letters,* 30(18): 1546–1548, September 1994.

26. M. J. M. Pelgrom, private communication.

27. M. Shoji, *CMOS Digital Circuit Technology,* Prentice-Hall, Englewood Cliffs, NJ, 1988.

28. H. B. Bakoglu, *Circuits, Interconnections, and Packaging for VLSI,* Addison-Wesley, Reading, MA, 1990.

29. M. Shoji, *Theory of CMOS Digital Circuits and Circuit Failures,* Princeton University Press, Princeton, NJ, 1992.

30. D. J. Allstot, S. Chee, S. Kiaei, and M. Shrivastawa, "Folded Source-Coupled Logic for Low-Noise Mixed-Signal IC's," *IEEE Transactions on Circuits and Systems—I,* 40: 553–563, September 1993.
31. J. A. Olmstead and S. Vulih, "Noise Problems in Mixed Analog-Digital Integrated Circuits," *Digest of the IEEE 1987 Custom Integrated Circuits Conference,* Portland, OR, pp. 659–662.
32. A. S. Moulton, "Laying the Power and Ground Wires on a VLSI Chip," *Proceedings of the 20th Design Automation Conference,* Miami Beach, FL, June 1983, pp. 745–755.
33. J. Trontelj, L. Trontelj, and G. Shenton, *Analog-Digital ASIC Design,* McGraw-Hill, London, 1989.
34. D. K. Su, M. J. Loinaz, S. Masui, and B. A. Wooley, "Experimental Results and Modeling Techniques for Substrate Noise in Mixed-Signal Integrated Circuits," *IEEE Journal of Solid-State Circuits,* 28: 420–430, April 1993.
35. B. R. Stanisic, N. K. Verghese, T. S. Rutenbar, L. R. Carley, and D. J. Allstot, "Addressing Substrate Coupling in Mixed-Mode IC's: Simulation and Power Distribution Synthesis," *IEEE Journal of Solid-State Circuits,* 29: 226–238, March 1994.
36. K. Joardar, "A Simple Approach to Modeling Cross-Talk in Integrated Circuits," *IEEE Journal of Solid-State Circuits,* 29: 1212–1220, October 1994.
37. T. J. Schmerbeck, "Minimizing Mixed-Signal Coupling and Interaction," *Proceedings of the European Solid-State Circuits Conference,* Ulm, Germany, pp. 28–37, 1994.
38. B. Abdi, "Mixed Signal IC Design for Data Communication," RF IC Design for Wireless Communications Systems Course, Lecture notes, Mead Microelectronics, Monterrey, CA, October 1994.
39. W. H. Hayt and J. E. Kemmerly, *Engineering Circuit Analysis,* 5th ed., McGraw-Hill, New York, 1993.
40. R. Lohia and A. Ali, "Parametric Formulation of CMOS Latch-up as a Function of Chip Layout Parameters," *IEEE Journal of Solid-State Circuits,* 23: 245–250, February 1988.
41. L. Herman, "Controlling CMOS Latch-up," *VLSI Design,* 6(4): 100–107, April 1985.
42. R. A. Rutenbar, "Analog Design Automation: Where Are We? Where Are We Going?" *Proceedings of the 1993 Custom Integrated Circuits Conference,* San Diego, CA, May 1993, pp. 13.1.1–13.1.8.

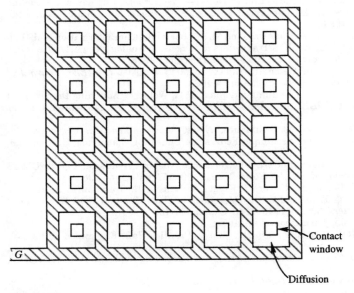

Contact window

Diffusion

Figure 6.27

Problems

Note: Unless otherwise noted, layout rules are the same as in Example 6.1. Electrical parameters are as in Examples 5.1 and 5.2.

6.1   Estimate W, L, and W/L for each device in Figs. 6.2 and 6.3. Assume the contact windows are 1 µm by 1 µm and that the lateral diffusion extent l_{lat} is 0.2 µm.

6.2   Suggest a way to estimate the effective W for the device geometries shown in Figs. 6.5b and c.

6.3   The idea behind the layout of Fig. 6.5c can be extended to large W/L ratios as shown in Fig. 6.27 [23]. The large squares are n^+ diffusion regions, and the small ones are contact windows. Assume the diffusion regions are denoted by a row number i and a column number j. The regions for which $i + j$ is even are to be part of the drain; the remaining regions are to be part of the source. Suggest a metallization pattern connecting all drain subregions in parallel and leading to a drain terminal; similarly, all source subregions should be connected to a source terminal. (*Hint:* The metal lines can run diagonally.) Assuming the contact windows shown are 1 µm on a side, estimate the values of W and L for the complete device. What are the advantages and drawbacks of this layout style, compared to the style shown in Fig. 6.3?

6.4   Using the layout rules of Example 6.1, lay out a resistor of 10 kΩ, using each of the approaches shown in Fig. 5.8. In each case, try to minimize the total layout area involved (including that of the shield, if present). Find the total layout area in each case.

6.5   Lay out the circuits shown in Fig. 6.28, following the layout rules of Example 6.1. The quantities next to the devices in the form a/b mean $W = a$ and $L = b$. Assume a lateral diffusion extent l_{lat} of 0.25 µm.

6.6   Draw the circuit implemented by the layout of Fig. 6.29, and give the sizes of all devices involved. Assume that the contact windows are 1 µm on a side and that the lateral diffusion extent l_{lat} (not shown) is 0.25 µm. (What you are asked to do in this problem is commonly known as *reverse engineering*.)

6.7   In a BiCMOS circuit, an *npn* transistor collector current must be 10 mA. To operate the device optimally (Sec. 5.3), it has been determined that its emitter current density (emitter current per unit of emitter area) must be about 100 µA/µm². Give a layout for the device, using the layout rules of Example 6.1. Assume that the maximum allowed current in the metallization layer is 1 mA/µm of width.

6.8   Consider two capacitors C_1 and C_2. Capacitor C_1 is square, with a side length equal to a. Capacitor C_2 is rectangular and equal to ηC_1, where $\eta > 1$. Show that the required lengths of the sides of C_2, so that the perimeter/area ratio is the same for the two capacitors, are as indicated in Fig. 6.12b.

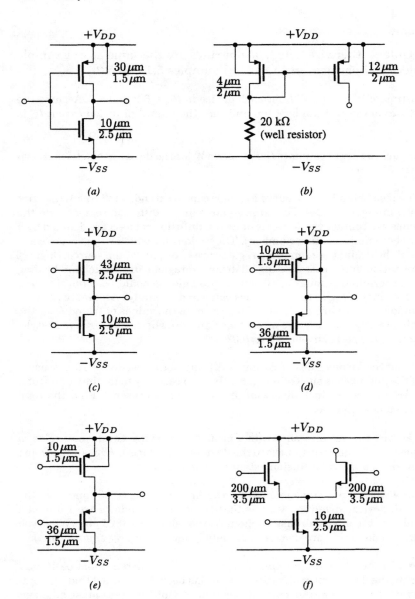

Figure 6.28

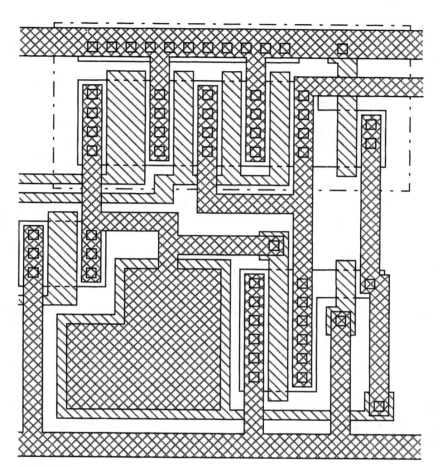

Figure 6.29

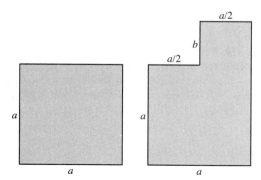

Figure 6.30

6.9 Consider two capacitors with top plates as shown in Fig. 6.30. Show that the perimeter/area ratio is the same in both, and determine the value of b required for a capacitance ratio $\eta > 1$.

6.10 Assume that in Fig. 6.13a there is nonzero oxide thickness gradient, as one goes horizontally across the capacitor array. Show that if this gradient is sufficiently small, the resulting variation of capacitance per unit area (with horizontal position) will be approximately linear. The oxide capacitance per unit area is inversely proportional to the oxide thickness.

6.11 Lay out two capacitors with ratios of (a) 4:1, (b) 4.5:1, and (c) 4.14:1. Show only the top plates for simplicity, in the style of Fig. 6.13.

6.12 Using the style illustrated in Fig. 6.17, lay out two capacitors with a ratio of 5:1.

6.13 Modify the layout of Fig. 6.19 so that the two devices connected in parallel are the side devices. Comment on the advantages and drawbacks of such a layout.

6.14 The ratio of the currents of two nMOS transistors, when identically biased, is to be 5:2 as accurately as possible. Give an appropriate layout.

6.15 Complete the layout in Fig. 6.20. Attempt to attain "perfect" matching between the two composite devices M1 and M2, even with respect to their parasitics. Comment on the advantages and drawbacks of your approach.

6.16 The W and L values of the device in Fig. 6.2d are exactly twice the corresponding values in Fig. 6.2c. Somebody proposes that the currents of the two devices can be accurately matched by biasing the two devices identically. Is this true? Why, or why not?

6.17 Two MOS transistors have been laid out identically for good matching. However, the suggestion of Fig. 6.21 was not followed, and the temperatures of the two devices differ by 5°C. Assuming the devices are biased with $V_{GS} - V_T = 0.5$ V, estimate their current mismatch (see Sec. 3.7).

6.18 The simple low-pass filter in Fig. 6.31 is made from a well resistor (Fig. 5.8f) with a width of 10 μm, in the fabrication process of Example 5.1. Due to digital switching elsewhere on the chip, the substrate in the vicinity of the resistor has 1-V high-frequency "noise" spikes (Sec. 6.5). Estimate roughly the

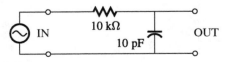

Figure 6.31

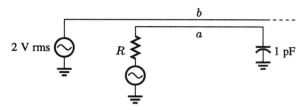

Figure 6.32

amplitude of the resulting noise spikes at the output of the filter. (*Hint:* Use the model of Fig. 5.10*b*, and do a worst-case calculation by making reasonable simplifications. Assume that the input voltage source is ideal.)

6.19   In Fig. 6.32, line *a* is 300 µm long and is fed by a signal source with an output resistance *R*. Line *b* runs parallel to line *a* at a minimum distance, and it carries a voltage signal of 2 V rms. The interline parasitic capacitance per unit of line length is 0.03 fF/µm. Estimate the signal crosstalk from line *b* to line *a*, assuming the resistance *R* is very large and that only capacitive effects are important. Suggest a way to reduce the crosstalk.

6.20   The ground line running from the pad to circuit *A* in Fig. 6.33 is 0.5 mm long. The fabrication process used is that in Example 5.1. The current of circuit *A* varies slowly with time between 0 and 1 mA.

(*a*) If the ground line is of minimum width (2 µm), what is the voltage interference caused by circuit *A* at point *X*?

(*b*) Assume that the above interference should be no more than 0.5 mV if circuit *B* is to function properly. What should the minimum width of the ground line be?

(*c*) How can the interference from circuit *A* to circuit *B* be reduced without a line-width increase?

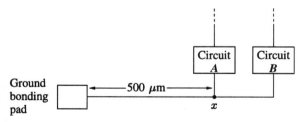

Figure 6.33

Additional MOS Transistor Modeling Information

We give below some expressions often used in MOS transistor models. The symbols used but not defined below are explained in Sec. 3.2. The device type is assumed to be nMOS. Further information and numerous references can be found elsewhere [1].

V_{T0} and ϕ_0 for Unimplanted Devices

A common approximation is

$$V_{T0} = V_{FB} + \phi_0 + \gamma \sqrt{\phi_0} \qquad (A.1)$$

where V_{FB} is the *flat-band voltage,* given by

$$V_{FB} = \phi_{MS} - \frac{Q_o'}{C_{ox}'} \qquad (A.2)$$

with ϕ_{MS} being the potential corresponding to the gate-body work function difference, Q_o' the effective semiconductor-insulator interface charge, and

$$\phi_0 \approx 2\phi_F \qquad (A.3)$$

where ϕ_F is the body's *Fermi potential,* given by

$$\phi_F = \phi_t \ln \frac{N_B}{n_i} \qquad (A.4)$$

with n_i the intrinsic carrier concentration (1.45×10^{10} cm^{-3} at room temperature).

These results are not valid for implanted devices. A "quick fix" often used for such devices, consisting of adding the implant charge per unit area to Q'_o in Eq. (A.2), does not work if the implant is not extremely shallow. The effective value of ϕ_0 for implanted devices can be smaller or larger than $2\phi_F$, depending on implantation details.

Built-in Potential for n^+p or p^+n Junctions

$$\phi_j \approx \frac{V_{\text{GAP}}}{2} + \phi_F \tag{A.5}$$

where V_{GAP} is the potential corresponding to the energy gap (1.12 V for silicon at room temperature) and ϕ_F is given by Eq. (A.4), with N_B being the doping concentration of the lightly doped side.

Zero-Bias Capacitance per Unit Area for n^+p or p^+n Junctions

The zero-bias capacitance per unit area is

$$C'_{jo} = \frac{\sqrt{2qe_sN_B}}{2\sqrt{\phi_j}} \tag{A.6}$$

with N_B the doping concentration of the more lightly doped side, ϕ_j as in Eq. (A.5), and the rest of the parameters as in Table 3.3. A rough estimate of C'_{jo} is

$$C'_{jo} \approx (0.3 \text{ fF/}\mu\text{m}^2) \sqrt{\frac{N_B}{10^{16} \text{ cm}^{-3}}} \tag{A.7}$$

Reference

1. Y. Tsividis, *Operation and Modeling of the MOS Transistor,* McGraw-Hill, New York, 1987.

B

A Set of Benchmark Tests for Evaluating MOSFET Models for Analog Design*

In various sections of this book (notably Secs. 3.12 and 4.9), readers were warned about problems encountered with several models found in popular simulators, when the models are used for analog design. We give below several benchmark tests to evaluate MOSFET models in this context. These have been found very useful over the years, as a necessary (but not sufficient) set of tests that a model should pass before we can begin to trust it for analog work. Ideally, the tests should be *quantitative* comparisons to measured data; however, even if such data are not available, one can get very useful indications by running just the simulations indicated, since they will at least show whether the model being tested gives correct *qualitative* behavior. Some examples of how well popular models fare when put to these tests will be given along the way, but we will avoid giving numerical comparisons, as we do not want to limit our comments to specific models with specific parameter values. All tests are to be done on nonminimum geometries (long and wide channels) at room temperature. Models that fail most of these tests include, e.g., Spice model levels 1, 2, and 3.

---

* This appendix is based on Y. Tsividis and K. Suyama, "MOSFET Modeling for Analog Circuit CAD: Problems and Prospects," *IEEE Journal of Solid-State Circuits,* 29:210–216, March 1994 (Copyright 1994 by IEEE).

Benchmark Test 0: Strong Inversion Current

This test concerns basic I-V characteristics accuracy; if this test fails, the model may be unsuitable even for digital circuit work. In strong inversion and with $V_{SB} = 0$, plot I_D versus V_{DS}, with V_{GS} as a parameter, for two devices: one with a small value, and one with a large value, of the body effect coefficient γ, but both with the same mobility-oxide capacitance product. Contrary to what is predicted by simple models often used for hand analysis, for the *same* V_{GS}-V_T values the two devices should behave significantly differently owing to their different values of α in Tables 3.1 and 3.2 [1]. Notably, the device with the large γ should exhibit lower $V_{DS,SAT}$ values and lower saturation current for the same V_{GS} value. For example, for $\gamma = 0.5$ V$^{1/2}$ the above quantities can be about 20 percent lower than those for a small-γ device.

Benchmark Test 1:
Weak and Moderate Inversion Current

For a V_{DS} value in the saturation region, plot I_D versus V_{GS}, with I_D on a *logarithmic* scale and including V_{GS} values well below the threshold. The shape should be as illustrated by the solid line in Fig. 3.10. Many popular models fail this test at least in the moderate inversion region, as shown by the broken line in the same figure.

Benchmark Test 2:
Transconductance/Current Ratio

Plot the transconductance/current ratio g_m/I_D (an important quantity for analog design) versus V_{GS} or versus $\log I_D$ (same ranges as for benchmark test 1). This is easy to do with simulators that allow numerical operations [e.g., with PSpice one can get I_D versus V_{GS} and then ask for a plot of $(dI_D/dV_{GS})/I_D$]; otherwise, the test can be tedious but will *still be worth running*, at least for V_{GS} values around kinks such as the one in Fig. 4.22a. Measurements give the type of shape shown by the solid line in Fig. 4.22b.* Several models give results of the type shown by the broken line in the same figure. A large error occurs in the moderate inversion region.

---

* As I_D is reduced, the quantity g_m/I_D peaks in weak inversion, but does not become exactly constant in it. This is due to minute deviations from exponential behavior for the current (too minute to be noticeable in plots like the one of Fig. 4.22a), which are predicted by detailed charge-sheet models [2, 1]. However, predicting a constant g_m/I_D in weak inversion would be acceptable for most purposes (except, of course, at such low current values that junction leakage becomes noticeable).

Benchmark Test 3:
Drain-Source Conductance

Plot g_{ds} (= dI_D/dV_{DS}) versus V_{DS} for a fixed V_{GS} value (or, better yet, obtain a family of such curves). The expected shape is shown in Fig. B.1a, but some models give the result in Fig. B.1b depending on the model parameter values. The reason is an unnatural transition from triode to saturation, shown in Fig. B.1c (see also Sec. 4.3.3). Other models do not produce such an abrupt change but *still* predict g_{ds} inaccurately in the transition from nonsaturation to saturation. Another problem in some models involves the expression used internally for g_{ds}. If the values of g_{ds} as a small-signal parameter are requested for several V_{DS} values (e.g., by asking for an operating point analysis at each such value) and plots of g_{ds} versus V_{DS} are generated, then a discontinuity or a sudden slope change can be observed at $V_{DS} = 0$. This is not in agreement with experiment and contradicts the behavior of dI_D/dV_{DS} obtained by differentiation on I_D-V_{DS} data obtained by using the same model, which show that nothing special should be happening at $V_{DS} = 0$.

Benchmark Test 4:
High-Frequency Transadmittance

Take the simplest possible model (*remove* all parameters having to do with parasitics, such as junction and overlap capacitances and series resistors). In the device statement, do *not* specify source or drain areas and perimeters. We suggest the above simplifications to make clear what the following problem is due to. Bias a 100-μm-long MOSFET in strong inversion saturation, where the intrinsic gate-drain capacitance is zero. Use an ac source in series with the gate bias, as shown in Fig. B.2a, and obtain a frequency response for the drain-ac current magnitude up to 10 GHz. Now break the device into two 50-μm-long devices, with their channels in series and with common gate and common substrate, and bias the combination as before. The combination should be equivalent to the single 100-μm device (remember, no junction area is supposed to exist at the intermediate point). Obtain the frequency response again. It should be the same as before. However, several common models give the behavior shown in Fig. B.2b. The behavior is totally different at high frequencies. This is the result of the fact that the models used do not take into account nonquasi-static effects (Sec. 4.6.2); the behavior predicted for the 100-μm single device is, of course, totally unreasonable.* This is wrong and contradicts both nonquasi-

---

* In fact, the use of *transcapacitors* in some models can produce even worse errors, predicting that the ac current magnitude goes *up* with frequency (this effect will only be seen if it is not masked by extrinsic capacitance effects).

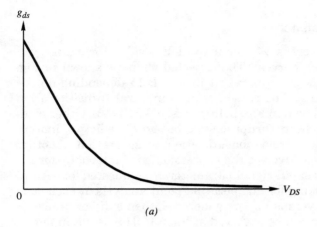

(a)

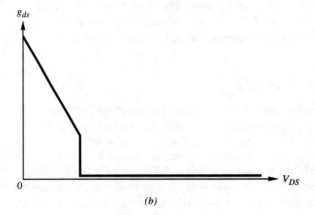

(b)

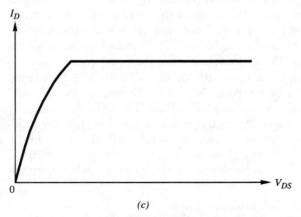

(c)

Figure B.1 g_{ds} versus V_{DS} for fixed V_{GS} and V_{SB}. (*a*) A qualitatively correct curve; (*b*) a plot using a popular Spice
model; (*c*) I_D versus V_{DS} for fixed V_{GS} and V_{SB} using the
same model.

static models and measurements [1, 3]. The two-device combination does a better job at approximating reality, since it is a two-element lumped approximation of what is actually a distributed channel effect. (In fact, such combinations, with two or more elements, can be used in lack of nonquasi-static models, for high-frequency small-signal work [1]; one should be careful, though, not to activate artificial short-channel effects in the subtransistors.) We note that nonquasi-static behavior has been experimentally demonstrated even in short-channel devices [3].

Benchmark Test 5: Thermal Noise

Bias a device with a fixed V_{GS} in strong inversion and at $V_{DS} = 0$ (by placing a zero-value dc current source between drain and source, as shown in Fig. B.3). Run a noise simulation, for a frequency low enough that the result is not affected by capacitances. Biased as indicated, the channel is equivalent to a resistor of value $R = 1/g_{ds}$, and it should show a thermal noise voltage with a power spectral density of $4kTR$ (e.g., 1.66×10^{-16} V^2/Hz for a g_{ds} of 10^{-4} A/V at room temperature). Many models give, depending on implementation, a value that is either a couple of orders of magnitude too low or even identically zero. The consequences are obvious for the design of circuits using MOSFETs as resistors.

Benchmark Test 6: 1/*f* Noise

Bias a device in strong inversion saturation, and run a noise simulation at frequencies where 1/*f* noise should be dominant. The noise current can be converted to a voltage across a 1-Ω resistor, placed in series

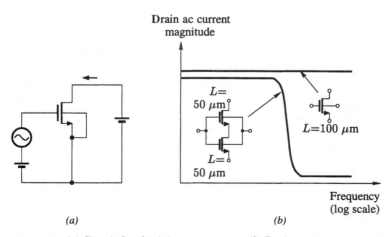

(a)   (b)

Figure B.2 (a) Circuit for obtaining ac response; (b) Drain ac current magnitude versus frequency for a 100-μm-long MOSFET and an equivalent combination of two 50-μm-long devices, using the level 2 Spice model.

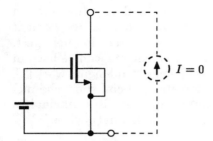

Figure B.3 Circuit for simulating thermal noise in the triode region.

with the drain (or, even better, a "noiseless resistor" implemented by using a self-dependent voltage-controlled current source). Now increase the channel area 10 times: Does the power spectral density of the equivalent input noise voltage (in square volts per hertz) decrease 10 times, as approximately observed in practice? Also, change V_{GS}; in most devices, the equivalent input noise voltage should be insensitive to this change.

Most models in use will fail some of or all the above tests. And, of course, even if they pass some tests qualitatively, the models still would have to pass them quantitatively, in comparison to measurements. Readers may want to take the results of the above tests to the model support team at their institutions or foundries.

Other Problems

The above benchmark tests cover only some of the major problems which, at present, seem to be present in most popular CAD models. Other problem areas include capacitances and noise in the moderate inversion region; transient response under nonquasi-static conditions; the influence of the body effect along the channel on thermal noise; noise at frequencies where nonquasi-static effects are observed; effective mobility dependence on V_{SB}; etc. Most of these problems are discussed in Ref. 1, where extensive references to the literature are given. The problems mentioned, as a rule, get worse for devices with short and/or narrow channels and for implanted (i.e., real!) devices. Also, particular models (and particular implementations of models in specific simulators) may have additional problems. Many circuit designers have formed a list of problems particular to the models/simulators they use.

Parameter Extraction

Even if a model could, in principle, do a decent job over certain bias ranges, often it is not given the opportunity to do so because of poor

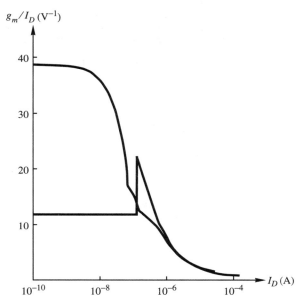

Figure B.4 g_m/I_D versus log I_D using the Spice level 2 and Bsim models with parameters as provided by a foundry service for the same fabrication process.

parameter extraction. As an example, we took the parameters provided by a well-known foundry service for models level 2 and Bsim, for the *same* fabrication process (obtained from measurements on the *same* devices). The I_D-V_{DS} curves in strong inversion (where the parameters were apparently extracted) gave relatively good agreement of one model with the other. However, a plot of g_m/I_D using the two models gave the results in Fig. B.4. The figure speaks for itself (note that no comparison of accuracy of the two models in weak inversion is implied here, since the foundry simply did not attempt to match the models to data in that region). Using the parameters from the same foundry also gave a six-order-of magnitude difference in $1/f$ noise predictions between the level 2 and Bsim models, a factor-of-3 discrepancy in the saturation g_{ds}, etc. Such problems are due to extraction being performed with only the digital designer in mind.

Many more benchmark tests can be found in Reference 4.

References

1. Y. P. Tsividis, *Operation and Modeling of the MOS Transistor,* McGraw-Hill, New York, 1987.
2. J. R. Brews, "A Charge Sheet Model for the MOSFET," *Solid-State Electronics,* 21: 345–355, 1978.

3. R. Singh, A. Juge, R. Joly, and G. Morin, "An Investigation into the Nonquasistatic Effects in MOS Devices with Wafer S-parameter Techniques," in *Proceedings of the International Conference on Microelectronic Test Structures,* Barcelona, Spain, March 1993.

4. Report, Sematech Compact Model Workshop, Sunnyvale, CA, August 11, 1995.

C

A Sample Spice Input File

We present here an example of an input file for the simulator Spice 2G and other equivalent Spice versions. A main purpose of this example is to illustrate the use of diodes for parasitic junctions and the use of scale factors. General information about Spice, and about the models in it, is provided elsewhere [1–4]. The reader is warned, however, that Spice model levels 1, 2, and 3, as well as many others, are in general inadequate for analog design (Secs. 3.12, 4.9, and 4.10 and App. B).

In Spice 2G, capacitance parameters for MOS transistor models are conveniently specified normalized to area or length. This practice, however, is not followed for models of bipolar transistors and diodes. For convenience, then, one can specify a fictitious *unit diode* (1-μm by 1-μm area) and a fictitious *unit transistor* (1-μm emitter width) and then can use an appropriate scale factor to specify the device connections.

Our circuit example is shown in Fig. C.1. The devices are assumed laid out as in Fig. C.2, and they are made using the process of Examples 5.1 and 5.2. The Spice input file is as follows:

```
CIRCUIT EXAMPLE
*
* Element Specification
*---------------------
*
vdd 1 0 dc 2.5
vss 0 2 dc 2.5
m1 1 1 3 2 mn w=10u 1=3u as=30p ad=30p ps=26u pd=26v
m2 2 2 4 4 mp w=20u 1=3u as=60p ad=60p ps=46u pd=26u
```

```
q1 3 0 4 2 qn 12
d1 2 4 jd 390
*
* Model Specification
*----------------------
*
* nmos model
*
.model mn nmos(level=2 ld=0.1e-6 vto=0.77 kp=77e-6 gamma=0.35
+phi=0.76 cj=0.31e-3 cjsw=0.17e-9 cgso=0.19e-9 cgdo=0.19e-9
+cgbo=0.3e-9 nsub=1.9e16 nfs=1e10 tox=200e-10 xj=0.14e-6
+ucrit=1.5e4 uexp=0.077 delta=2 kf=4e-13 af=1.2)
*
* pmos model
*
.model mp pmos(level=2 ld=0.21e-6 vto=-0.87 kp=28e-6 gamma=0.70
+phi=0.72 cj=0.62e-3 cjsw=0.26e-9 cgso=0.42e-9 cgdo=0.42e-9
+cgbo=0.3e-9 nsub=8.8e16 nfs=1e10 tox=200e-10 xj=0.3e-6
+ucrit=1.2e4 uexp=0.063 delta=2 kf=7e-15 af=1.2)
*
* well-substrate junction model (for a fictitious well - 1 um X
1 um area!)
*
.model jd d(is=2e-16 cjo=0.28e-15 vj=0.8)
*
* npn bipolar transistor model (for a fictitious, 1 um emitter
width!)
*
.model qn npn(is=1e-18 bf=90 vaf=75 ikf=0.005 ise=3e-17 ne=2
+var=6 rb=420 rbm=230 re=27 rc=54 cje=8e-15 vje=0.9 mje=0.33
+cjc=6e-15 vjc=0.55 mjc=0.3 cjs=18e-15 vjs=0.6 mjs=0.4
+tf=15e-12 xtf=15 vtf=3 itf=16e-3 tr=2e-8 xtb=0.9 kf=3e-15 af=1)
*
* Analysis Commands
* -----------------
*
.op
*
.end
```

In the MOSFET element specifications, L is the drawn channel length; i.e., it is the quantity shown as $L + 2l_{\text{lat}}$ in Fig. 6.1. Thus, all effective channel lengths (denoted simply as L in this book) must be augmented by $2l_{\text{lat}}$ before they are entered into the input file. The quantities as and ad are the source and drain areas in square meters; ps and pd are the source and drain perimeters in meters. The prefix "u" stands for 10^{-6}; the prefix "p" stands for 10^{-12}.

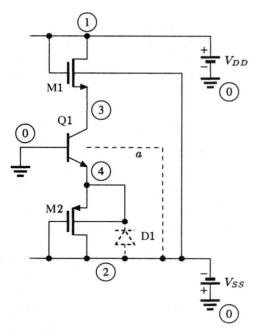

Figure C.1

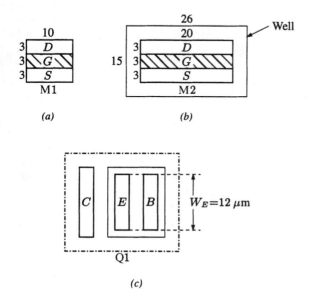

(a)   *(b)*

(c)

Figure C.2

The well-substrate junction for the pMOS device is specified as a separate diode ($d1$ in the file). The *model jd* statement is for a fictitious *unit well* (1 μm by 1 μm). Thus, for the 26-μm by 15-μm well of device M2 (Fig. C.2b), a scale factor of $(26 \times 15)/(1 \times 1) = 390$ is included in the $d1$ element specification (sidewall area is ignored for simplicity; if desired, it can be included by adding an extra diode and an appropriate model).

The *model qn* statement is for a fictitious *unit width* device ($W_E = 1$ μm in Fig. 5.26). Thus, for a device with $W_E = 12$ μm (Fig. C.2), a scale factor of $12/1 = 12$ is specified in the $q1$ element specification. (The simple W_E-scaling idea discussed in Sec. 5.3 is used for simplicity.)

The fourth terminal in the specification of the bipolar transistor Q1 is needed for modeling the collector-substrate junction (Fig. 5.14). This fourth terminal is indicated by broken line a in the schematic. Spice will automatically include a collector-substrate junction; no separate diode should be included in the file for this purpose.

If the circuit contained more than one pMOS device in separate wells, one would still need only one diode model statement, as long as the proper scale factor was included in each of the diode element statements. Also, if more npn bipolar transistors were present and if they were all laid out in the style of Fig. C.2c (with only W_E being possibly different), one would still need only one bipolar transistor model; of course, the appropriate scale factor should be included in each bipolar transistor element statement.

To help its algorithms converge, Spice adds large resistors (default value 1e12 ohm) in parallel with every pn junction. When the simulation of leakage currents is important, this should be taken into account in the interpretation of the results. This and other default values can be modified by the user [3].

One should not seek an exact correspondence between the parameters in Spice models and those in the simple models in this book. Seemingly "corresponding" parameters may be treated differently in each type of model and may have to be assigned different numerical values in order to model the same device. Even then, it may happen that no satisfactory numerical agreement can be obtained between the predictions of Spice and those of hand analysis. The reader is urged to analyze by hand the circuit of Fig. C.1, using the simple models in this book (with parameters as in Examples 5.1 and 5.2) and to simulate it by using the above file in Spice. A detailed comparison of the results obtained will be instructive.

References

1. L. Nagel, *SPICE2: A Computer Program to Simulate Semiconductor Circuits,* ERL Memorandum ERL-M520, Electronics Research Laboratory, University of California, Berkeley, May 1975.

2. A. Vladimirescu and S. Liu, *The Simulation of MOS Integrated Circuits Using SPICE2*, ERL Memorandum UCB/ERL-M80/7, Electronics Research Laboratory, University of California, Berkeley, February 1980.
3. P. T. Tuinenga, *SPICE—A Guide to Circuit Simulation and Analysis Using PSpice*, Prentice-Hall, Englewood Cliffs, NJ, 1992.
4. P. Antognetti and G. Massobrio, *Semiconductor Device Modeling with Spice*, McGraw-Hill, New York, 1993.

Index

ABOUT THE AUTHOR

Yannis Tsividis is Professor of Electrical Engineering at
Columbia University. He has previously taught at the
University of California, Berkeley; MIT; and the National
Technical University of Athens. Professor Tsividis has
also worked at AT&T Bell Laboratories and Motorola
Semiconductor, and he has consulted on mixed analog-
digital chips for a variety of companies. He is the author of
Operation and Modeling of the MOS Transistor, published
by McGraw-Hill.